T0335874

THE AGE OF THE EARTH

A Physicist's Odyssey

THE AGE OF THE EARTH

A Physicist's Odyssey

Archibald W. Hendry

World Scientific

NEW JERSEY · LONDON · SINGAPORE · BEIJING · SHANGHAI · HONG KONG · TAIPEI · CHENNAI · TOKYO

Published by

World Scientific Publishing Co. Pte. Ltd.
5 Toh Tuck Link, Singapore 596224
USA office: 27 Warren Street, Suite 401-402, Hackensack, NJ 07601
UK office: 57 Shelton Street, Covent Garden, London WC2H 9HE

British Library Cataloguing-in-Publication Data
A catalogue record for this book is available from the British Library.

THE AGE OF THE EARTH
A Physicist's Odyssey

ISBN 978-981-3279-69-8
ISBN 978-981-120-131-8 (pbk)

For any available supplementary material, please visit
https://www.worldscientific.com/worldscibooks/10.1142/11260#t=suppl

Desk Editor: Nur Syarfeena Binte Mohd Fauzi

Typeset by Stallion Press
Email: enquiries@stallionpress.com

Printed in Singapore

For Muriel

Contents

Chapter 6. Concluding Remarks   123

Index   129

Preface

The age of the Earth. It's one of the most contentious issues around. There's no middle ground. The antagonists are implacable. One group insists the Earth is only about 6000 years old. The other says it's billions of years old.

So how is it that, in this day and age, we can have two totally different answers? Where do they come from?

I've often wondered about this. Every now and then, I'd start to take a closer look, but never quite finished the job. Like most people, kept putting it off. Saying I'll finish it — later. Find out *exactly* how the age of the Earth was figured out. Not just the story — that's been told many times by lots of people. But the real nuts and bolts. No holds barred.

This book is the outcome of these investigations.

Previously,[1] I've written about the major scientific efforts that led to a greater understanding of *planetary orbits* — in particular, the insights of Ptolemy, Copernicus, and Kepler. Their struggles led eventually to the discovery of the famous $1/r^2$ law of *gravity* by Isaac Newton.

Here, I'm going to take a look at how several other great people determined — or thought they had determined — how old the Earth is. These include James Ussher (the famous 17th-century biblical scholar), James Hutton (the founder of modern geology),

---

[1]Hendry, Arch. W. 2018. *A Physicist's Odyssey ∼ from orbits to gravity*. Witney, Oxon: Robt. Boyd Publ., available from amazon.co.uk.

Charles Darwin (of *Origin of Species* fame — or in the eyes of some, shame), and William Thomson (alias Lord Kelvin, the great mathematical physicist of the 19th century). Finally, I'll examine what modern physics has to say.

The answers are all different. Which one — if any — is correct?

Nuts and bolts of course to a theoretical physicist like me means that, sooner or later, there's going to be some mathematics. And it's unavoidable. Especially when it comes to Kelvin. His achievement was to *calculate* the age of the Earth. Based on the physics and mathematics of his day. So we must do the same. Which of course would lead to the text being overburdened with too many formulae and equations. So I've tried to avoid this, keeping the mathematics in the text to a minimum (which is not zero!), putting the heavy-duty mathematics into Appendices. Hopefully, in the text, there are enough words between the equations to indicate where all the answers come from.

Our goal then is to work out and understand how the various determinations of the age of the Earth were done. Might get a better appreciation of both sides of the argument. Which would be good for *everyone* to know! Not just me.

Acknowledgments

I'd like to thank several friends who helped me through this particular journey: James Ackerman, Professor of Religious Studies, who long ago helped me appreciate better the Hebrew *Masoretic Bible*; Rev. Beverley Tasker who helped me understand more clearly the biblical Creation stories; Ms. Naomi Cohenour for her continued valiant efforts at converting my manuscript into a pdf file; and Robert Boyd for his friendly expert support.

Nur Syarfeena Binte Mohd Fauzi and Christopher Davis were instrumental in guiding this book smoothly through World Scientific Publishing Co. I thank them both.

Lastly, I thank my wife Muriel for her continued love and support throughout this second odyssey.

Chapter 1

The "Biblical" Age of the Earth

"...all History, as well sacred, as Prophane, and Methodically digested..."

Six thousand years. It's become famous. That's the age of the Earth. At least according to some people. They say it's based entirely on the Bible. It's all there. Just add up the numbers.

Trouble is — that's easier said than done. Have you ever come across anyone who *has* added up the numbers? Doubt it. I haven't. And that includes quite a few people I have known over the years who firmly believe in this 6000 years.

Have you tried to do it yourself? How does this number of 6000 years come about? It's got to be all there in the Bible. Right? A little detective work should be all that's necessary. So why don't we just do it?

The Bible of course is no quick read. There are words — lots of them. I reckon about 800,000 or so. And numbers. That's what we're after. Put them together. Form a biblical chronology.

Over the centuries, lots of famous people have done exactly this. Among them, the great astronomer Johannes Kepler (1571–1630) and the incomparable mathematician Isaac Newton (1643–1727). Definitely two people who knew how to count! And their final answers? Both arrived at numbers close to 6000.

But geniuses though they were, neither Kepler nor Newton is the person responsible for the most famous chronology of all. That honor goes to one, James Ussher (1581–1656). Who? Never heard of him? He's long since forgotten. But he's someone who, toward the end of

1

his life, was held in such high esteem that when he died, England's Lord Protector himself, Oliver Cromwell, decreed that Ussher's body be laid to rest, with full honors, in Westminster Abbey. Beside the other great worthies of the land.

So how did it come about that James Ussher's numbers became so famous? And accepted? Perhaps surprisingly, it involved a young enterprising publisher in London by the name of Thomas Guy.

Apparently, around 1675, Guy had a brainwave. Combine Ussher's dates (published in the 1650's) with the Authorized Version of the Bible (published in 1611), the King James Version. Put the dates at the top of each page. Dates and text together. What a powerful combination. And as the King James Bible became more widely accepted (at least in the English-speaking world), so too did Ussher's dates. (The Anglican Church gave its official blessing in 1701.) Soon everyone knew that the world had been created in the year 4004 BC. No doubt about it. There it was in black and white. On page one, Genesis 1.

And what about the date for that cataclysmic event that everybody knows about — Noah's Flood? It's at the top of the page for Genesis 7: the year 2348 BC, only 1500 years or so after Adam had taken that fateful bite from the forbidden fruit. (Sorry, "apple" never mentioned.) Now everything had a date.

Like James Ussher, Thomas Guy is barely remembered nowadays. But his presence is still invisibly there. For he used the profits from his publishing business and subsequent investments for a worthy cause — founding a much-needed new hospital in London in 1721. Which, in its modernized form, is still there — the world-famous Guy's Hospital. Near Westminster Bridge.

Over the centuries, the King James Bible went from strength to strength, even though its language belonged to Shakespearean times. Revered by vast millions of people, even today. Its words are so powerful, so evocative. In places, it sounds like God himself is speaking. Maybe it's not so surprising that some people have come to believe that the words themselves *are* God's very own.

And likewise for those dates at the top of the pages. So, if it says that Creation took place in 4004 BC, clearly the world is now about

6000 years old. Can't argue with that. But we can do some checking. Or at least find out where Ussher got all these dates.

The numbers beckon us!

(There's more on the King James Version of the Bible in Appendix 1; and on Archbishop Ussher in Appendix 2.)

Biblical chronology

Where to begin? No better place than the Beginning, Genesis 1. There, it tells us that God created the Heaven and the Earth. In six days. On the sixth day itself, God created mankind, both male and female.

And so the chronology starts. From Adam. It will lead us through his many descendants, their wanderings in the wilderness, years of trials and tribulations, of captivity and freedom; then judges and kings. This is the path we follow. One step at a time.

From Adam to Noah

Fortunately, the Old Testament gets us off to a good start. Genesis 5 begins with the words, "This is the book of the generations of Adam", and proceeds to list the first ten generations. It takes us from Adam to Noah, with verses like, "And Adam lived a hundred and thirty years, and begat a son...Seth", and "And all the days that Adam lived were nine hundred and thirty years". All the numbers we need (and more) are there.

The most important numbers among the many are the ages of the fathers when their (successor) sons were born. From these, we can build up a continuous chronology. Accepting the numbers as given. (Repeat — taking the numbers as given.)

The various names and ages for these first ten generations are listed in Table 1.

It's Genesis 7:6 and 11 that pinpoint the date of that great event, the Flood. We are told that Noah was 600 years old when it all happened. Since Noah was born 1056 years after Creation (Table 1), the almighty Flood took place 1656 years after Creation. Given that Ussher later determined that Creation occurred in the year 4004

Table 1: From Adam to Noah.

Name	Year of birth since Creation	Age when son was born	Age at death	Year of death
Adam	0	130	930	930
Seth	130	105	912	1042
Enos	235	90	905	1140
Cainan	325	70	910	1235
Mahalalel	395	65	895	1290
Jared	460	162	962	1422
Enoch	622	65	365	987
Methuselah	687	187	969	1656
Lemach	874	182	773	1651
Noah	1056	502	950	2006

BC, this means that the Flood took place in the year 2348 BC. And indeed, that is the date at the top of the page for Genesis 7 in my (old) copy of the King James Version of the Bible.

But wait. Not so fast. A little problem. There's a slight ambiguity about how old Noah was when his (successor) son Shem was born. Let's see.

Take Genesis 5:32. It reads "And Noah was five hundred years old, and Noah begat Shem, Ham and Japheth". Three sons. The same age? That's what a literal reading of this verse could mean. Which would imply that Shem would be 100 years old when the Flood occurred.

Later verses (such as Genesis 9:24 and 10:21) however seem to indicate that the brothers were not all of the same age. Moreover, Genesis 11:10 reads, "Shem was one hundred years old, and begat Arphaxad two years after the Flood". Which could imply that Shem was only 98 years at the time of the Flood (thus being 100 years old two years later when Arphaxad was born). Which in turn implies that Noah begat Shem when he was 502 years old (not 500 years).

Different answers. Admittedly not very much. Two years. Still, different. Which one is correct? Ussher decided on the *second* version (which is the one indicated in Table 1). It must have occurred to Ussher that dates weren't quite so pinned down as he might have hoped. A foretaste of things to come?

A couple of additional comments:

- Chronologies are text specific. Ussher followed the Hebrew *Masoretic Text*. The numbers in other texts — such as the Latin *Septuagint* — are all different. Leads to different dates.
- The King James Version of the Bible (borrowing from Tyndale) likewise followed the *Masoretic Text*. Thomas Guy may not have been fully aware of it, but Ussher and the King James were a perfect match!

From Noah to Abram

Now for the next step — the generations of Shem. Another batch of ten. All the relevant ages can be deduced from Genesis 11:10-32, and are listed in Table 2.

But wait. Another little problem. Another ambiguity? It involves Abram and his father Terah.

Genesis 11:26 states unequivocally that "Terah lived seventy years, and begat Abram...". The family was living at this time in the city of Ur, located in Lower Mesopotamia. Genesis 11:31 and 32 go on to say that Terah later moved with his family to the city of Haran, which lies on the Euphrates river in Upper Mesopotamia; Terah died there, 205 years old.

Table 2: From Shem to Abram.

Name	Year of birth since Creation	Age when son was born	Age at death	Year of death
Shem	1558	100	600	2158
Arphaxad	1658	35	438	2096
Shelah	1693	30	433	2126
Eber	1723	34	464	2187
Peleg	1757	30	239	1996
Reu	1787	32	239	2026
Serug	1819	30	230	2049
Nahor	1849	29	148	1997
Terah	1878	70	205	2083
Abram	1948	100	175	2123

The story continues in Genesis 12:1-5 where God calls on Abram to leave Haran. Which Abram does with his family and heads for Canaan. Verse 4 specifically tells us that Abram was 75 years old at the time.

Seems straightforward. Terah is living in Ur with his family; Abram born there when Terah is 70 years old. They move to Haran. When Abram is 75 years old, he and his family depart Haran, leaving his father to die there when he was 205 years old.

But Ussher had a problem with this interpretation. He seemed to think that the order in which the story is told was important. In particular, God's call to Abram and his subsequent departure from Haran is mentioned only *after* we are told of Terah's death. So, Ussher had Abram stay in Haran while his father Terah was alive. A close-knit family. Only after Terah's death did Ussher have Abram leave for Canaan — at age 75.

This changes the numbers. It now implies that Terah was aged 130 when Abram was born (clearly contradicting Genesis 11:26), a shift of 60 years.

Ussher's revised numbers for Terah and Abram are given in Table 3. It shows that Abram was born in the year 2008 after Creation.

The corresponding timeline from Adam to Abram is shown in Figure 1.

From Haran to Egypt

Abram is now on the move, from Haran in Upper Mesopotamia along the Fertile Crescent into the land of Canaan. According to Ussher, Abram started this journey the year his father died; he was 75 years old.

Table 3:   Revision of dates for Terah and Abram.

Name	Year of birth since Creation	Age when son was born	Age at death	Year of death
Terah	1878	130	205	2083
Abram	2008	100	175	2183

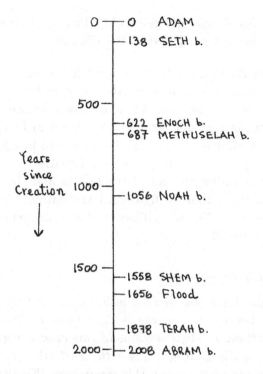

Figure 1: Timeline from Adam to Abram.

The story continues as follows. In Canaan, when he was 99, God made a covenant with Abram — thereafter to be known as Abraham; and in succeeding chapters of Genesis, we follow the lives of the other patriarchs Isaac and Jacob. Then Joseph, who was sold by his brothers and taken to Egypt. Eventually, as a result of famine, Jacob is forced to move with his remaining family to the Land of Goshen in Northern Egypt; there they settled under the protection of Joseph, by that time the Pharaoh's right-hand man.

The year of the family's move to Egypt turns out to be a key date in biblical chronology, as we'll see shortly. The specific information we need is the following:

Genesis 21:5   Abram was 100 years old when Isaac was born.

Genesis 25:26   Isaac was 60 years old when Jacob was born.

Genesis 47:9   Jacob was 130 years old when, on arrival in Egypt,
he was presented to the Pharaoh.

Assuming (with Ussher) that Abram left Haran when he was 75,
the above information tells us that Abram spent twenty-five years in
Canaan before Isaac was born. Thus, the total length of time spent
by the Israelites in Canaan before they settled in Egypt was (25 +
60 + 130) years, that is, 215 years. Their arrival in Egypt therefore
was (2008 + 75 + 215) = 2298 years after Creation.

We've now come to the end of the begets and the begats. Most
people seem to think that, to figure out the date of Creation, that's
all you need to know. Not true. They take you only so far. And we've
still a long way to go!

Sojourn in Egypt

The Israelites have now arrived in Egypt. How long did they stay
there? Lo and behold, the answer is right there in Exodus 12:40.

Trouble is, there are two versions of this crucial verse!

According to the Hebrew *Masoretic Text* (the one that Ussher
had been following up till now), this verse says, "Now the sojourning
of the children of Israel, who dwelt in Egypt, was four hundred and
thirty years".

The *Septuagint* on the other hand says that the time the Israelites
"dwelt in the land of Egypt *and* in the land of Canaan was four
hundred and thirty years" (my italics). Since we have just figured
out that the Israelites spent two hundred and fifteen years in Canaan
before entering Egypt, the *Septuagint* implies that they spent only
an additional two hundred and fifteen years in Egypt itself, not four
hundred and thirty.

Decisions, decisions. Which one to take?

Although Ussher had followed the *Masoretic Text* so far, on
this one he decided to go with the *Septuagint*. His choice is not
unreasonable. The Israelites had been in Canaan for about four generations (Abraham, Isaac, Jacob, and Joseph, say). Likewise, for their
stay in Egypt (Levi, Kohath, Amram, and Moses, say). So, you'd

expect they would spend similar amounts of time in the two places. That their stay in Egypt was exactly the same as their stay in Canaan (two hundred and fifteen years) is of course a bit of a miracle. But then, miracles are the stuff of the Bible!

Ussher's choice was the symmetric one — equal times in Canaan and in Egypt. Which gives us for the year of the Exodus (2,298 + 215) = 2,513 years after Creation.

Exodus to the fourth year of Solomon's reign

This seems to be an extraordinarily big jump — from the Israelites leaving Egypt to their conquest of Canaan and settlement there, then ruled by a series of judges and kings. Certainly not an easy task for any biblical chronologist — there are so many complications and unknowns. But amazingly, 1 Kings 6:1 tells it all: "And it came to pass in the four hundred and eightieth year after the children of Israel came out of the land of Egypt, in the fourth year of Solomon's reign over Israel...that he began to build the house of the Lord".

How lucky we are to have this overarching number! Especially since, on examining the biblical record, there aren't enough details to figure it all out bit by bit.

Let's take a look at that. Here are the major steps, with years provided when given.

1. The Israelites, led by Moses, wander in the wilderness for forty years (Numbers 14:33-34). (As we'll see, that number 40 keeps appearing time and again throughout the Old Testament — we've already met Noah's Flood which lasted forty days and forty nights. Likewise, in the New Testament — for example, Jesus was tempted in the wilderness for forty days.)

2. The Israelites, now led by Joshua, engage in the partial conquest and settlement of Canaan. We are not told how long he was their leader (though in Joshua 24:29, we are told that he died at the age of 110).

3. Then comes a long period in which the Israelites were governed by a series of judges (Judges 3–6), see Table 4.

Table 4: The Judges.

Years of oppression	Judge	Years of peace
8	Othniel	40
18	Ehud	80
	Shamgar	
20	Deborah	40
7	Gideon	40
3	Abimelech	
	(usurper)	
	Tola	23
	Jair	22
18	Jephthah	6
	Ibzan	7
	Elon	10
	Abdon	8
40	Samson	20

The Book of Judges seems to record a series of cycles, each cycle involving a period of oppression, the appointment of a new judge, then a period of peace. Adding together the number of years given for the various cycles, we get a total of 410 years. That's a big fraction of the 480 years allotted for the entire stretch from the Exodus to the fourth year of King Solomon's reign. Experts take it to imply that the judges did not govern sequentially. More likely, they ruled regionally (just like England before the Norman Conquest in 1066 AD, when several kingdoms such as Wessex, Mercia, Northumbria, etc. co-existed) and were overlapping time-wise.

4. The two priests Eli and Samuel. We are told (1 Samuel 4:18) that Eli was a priest for 40 years (that number again), but it is not recorded how long Samuel served from the time of Eli's death until Saul was appointed king.

5. Saul's reign. Oops! It's incredible. The Hebrew text for 1 Samuel 13:1 reads, "Saul was (blank) years old when he began to reign; and he reigned (blank) and two years over Israel". The actual numbers are missing! Okay, what does it say in the *Septuagint*? That might tell us. Afraid not — the whole verse is missing!

So how do we know how long Saul reigned? Somewhat of a puzzle, we find the answer in the New Testament (Acts 13:21): Saul reigned for 40 years.

6. David's reign. No problem here. 2 Samuel 5:4 tells us explicitly that David was 30 years old when he began to reign; he reigned for (you guessed it) 40 years.

7. The first four years of Solomon's reign. No surprise how long Solomon's entire reign was. It's there in 1 Kings 11:42. *Déjà vu* — 40 years.

Just adding together the numbers we know about gives many more years than 480. But there are large uncertainties. In his chronology, Ussher stayed with the overarching number of 480 years.

According to Ussher then, the building of Solomon's Temple was started in the year (2513 + 480), that is in the year 2993 after Creation.

Doesn't sound particularly significant — 2993. But it is — maybe. The reason is this. Revealed in 1 Kings 6:38. Which says that Solomon's Temple took seven years to build. That would be in the year 3000 after Creation. A most significant number for many people, including Ussher.

Why was this? It goes way back many centuries before Ussher, a feeling that maybe the history of the world went through distinct eras, each lasting a 1000 years. Where does that come from? The Bible, of course. Take, for example, 2 Peter 3:8 which says that "one day is with the Lord as a thousand years". But Genesis 1 tells us that God made the world in six days. Could a God-associated day correspond to a 1000 years of human experience? Many ecclesiastics believed this to be the case. And so eras each lasting a 1000 years were expected, at least by some. The fact that God's Holy Temple — the very first one for God's Chosen People — was completed in exactly 3000 years after Creation was significant to many. And it no doubt gave encouragement to Ussher that his chronology was on the right track.

Figure 2 continues our time-line up to the year 3000 after Creation.

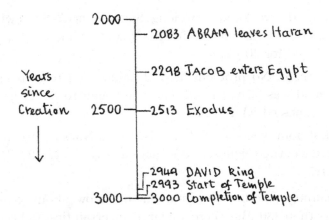

Figure 2: Timeline from Abram to the Temple.

The Temple: from construction to destruction

The story continues, but the counting is getting harder!

At this stage, we start with the remaining 36 years of Solomon's reign. But on Solomon's death, the nation is split in two — a northern kingdom retaining the name Israel with the city of Samaria eventually becoming its capital; and a southern kingdom which takes the name Judah, with its capital at Jerusalem.

Thereafter, a long line of kings for each country, specified in 1,2 Samuel and 1,2 Kings (see Table 5).

According to the numbers given, the northern kingdom lasted 241 years. It came to its end with the destruction of Samaria by the Assyrian king Sargon II and the dispersal of many of its people throughout Assyria.

The southern kingdom (through which David's line continued) fared better, lasting 394 years. It met its end with the destruction of Jerusalem by the Babylonian king Nebuchadnezzar II and the deportation of many of its citizens to Babylon.

It's a complicated history. And unfortunately, unlike the previous section, there's no overarching number to tell us how long this period lasted.

So, what shall we do? Let's start by adding the relevant numbers together, from the start of the building of the Temple, that is — the

Table 5: The Kings of Judah and Israel.

Judah (southern kingdom)		Israel (northern kingdom)	
King	**Years**	**King**	**Years**
Rehoboam	17	Jeroboam I	22
Abijam	3	Nadab	2
Asa	41	Baasha	24
Jehoshaphat	25	Elah	2
Jehoram	8	Zimri	(7 days)
Ahaziah	1	Omri	12
Athaliah	7	Ahab	22
Jehoash	40	Ahaziah	2
Amaziah	29	Jehoram	12
Azariah (Uzziah)	52	Jehu	28
Jotham	16	Jehoahaz	17
Ahaz	16	Joash	16
Hezekiah	29	Jeroboam II	41
Manasseh	55	Zechariah	(6 mon.)
Amon	2	Shallum	(1 mo.)
Josiah	31	Menahem	10
Jehoahaz	(3 mo.)	Pekaniah	2
Jehoiakim	11	Peka	20
Jehoiachim	(3 mo.)	Hoshea	9
Zedekiah	11	**Destruction of Samaria**	
Destruction of Jerusalem			

remaining 36 years of Solomon's reign plus the years of all the kings of Judah. What do we get? Lo and behold, the sum total is 430 years! That's a number we've met before — the time the Israelites spent in Canaan plus the time they spent in Egypt. Was this a coincidence? (Or were the authors of the Old Testament just fascinated — like physicists — by the idea of symmetry?)

Apparently, Ussher must have thought so. Previously, we had to reduce the time during which Israel was governed by the judges to less than the total number of years found by blindly adding together the numbers that are assigned to them individually (Table 4). Perhaps there was a similar situation for the kings of Judah listed in Table 5. After all, there may have been overlapping reigns or co-regencies. Most likely then, the sum total of 430 years was an over-estimate. In the end, Ussher reduced that number slightly to 423.

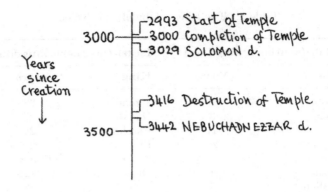

Figure 3: Timeline from the Temple to Nebuchadnezzar.

Ussher's date then for the destruction of Solomon's Temple in Jerusalem by the Babylonians was (2993 + 423) or 3416 years after Creation, as shown in the timeline in Figure 3.

The Exile and beyond

If it was becoming difficult to put specific dates to biblical events, now it becomes impossible. But how could that be? Why was Ussher so certain? How did he get his numbers beyond the Exile? Here is a tale that few people seem to know about.

We've now arrived at the times of Ezra and Nehemiah. Wonderful stories to read. But what about our chronology? Where's the information? Surely the numbers are there, somewhere. Sorry, no such luck.

Yes, some precise details are given. Such as the captives in Babylon being given the opportunity to return to Jerusalem in the first year of the reign of King Cyrus of Persia (Ezra 1:1-3). And the completion of the rebuilding of the Temple in Jerusalem in the sixth year of the reign of King Darius (Ezra 6:15). Fine, but we're not told, for example, how many years there are between these two important events. And about the stay in Babylon itself — how long did that actually last?

You might think that such important information would have been recorded somewhere in these Books. But no, it isn't there.

There's a period of 70 years mentioned in several places (2 Chronicles 36:21, Jeremiah 25:11-12 and 29:10, Zechariah 1:12), but it's uncertain what it refers to exactly — the length of the rule of the Babylonian Empire? The time between the destruction of the Temple and the completion of it being rebuilt? All very puzzling and inconclusive.

In fact, biblical chronology as such has come to an end. What? Yes, the important information we are looking for in the Old Testament just isn't there. And that's followed by an unspecified gap between the Old and New Testaments.

So what was Ussher to do? How could he continue to put it all together? How was he able to complete his chronology all the way from Creation through to the birth of Jesus and beyond?

The linkage and the 6000 years

In spite of what is often believed, Ussher did *not* deduce the dates of his biblical chronology *solely* from the Bible! Nor did he ever claim that he had. If only people had taken a closer look at the title page of his *Annales* — the English translation of which was published in 1658. That makes it perfectly clear. It says there that the *Annales* are based on "all History, as well Sacred, as Prophane, and Methodically digested, by the most Reverend JAMES USSHER".

There was of course another chronology available, even in the 17th century. It was a chronology based on ancient writings, historical records, and archaeology. The idea there was to start in more modern times — say the time of Jesus — and work backward. Through the Roman period, the Greeks, the Persians, the Medes, the Babylonians, the Assyrians, and so on. (Incredibly, Ussher taught himself many of these ancient languages, just to be able to read everything for himself — he didn't trust other people's translations.) And historians had been able to fit together the pieces of the historical jigsaw and come up with specific dates.

Clearly, these two chronologies — Ussher's biblical chronology marching forward from Creation toward the birth of Jesus, and the historical chronology reaching back in time from the birth of

Jesus — had to overlap sooner or later. All Ussher needed was an event that occurred in both his biblical chronology and in this historical chronology. Matching the two dates would then allow him to make the connection.

So what was Ussher's crucial event? His cross-link?

Ussher found it in the death of the Babylonian king Nebuchadnezzar II. There, in 2 Kings 25:27, was the clue he was looking for: "And it came to pass in the seven and thirtieth year of the captivity of Jehoiachin king of Judah... that Evilmerodach king of Babylon in the year that he began to reign did lift up the head of Jehoiachin king of Judah out of prison...".

This Evilmerodach was the son of Nebuchadnezzar, who had become king of Babylon on his father's death. So apparently Nebuchadnezzar had died in the 37th year of Jehoiachin's captivity. What had happened earlier was that, shortly after Jehoiachin had become king of Judah (2 Kings 24:8), in the year 3405 after Creation according to Ussher, Nebuchadnezzar had subdued Jerusalem, taking Jehoiachin prisoner and deporting him back to Babylon. Apparently, he was still there in prison 37 years later when Nebuchadnezzar died. (Evilmerodach subsequently released him, but kept him at the Babylonian court.)

So, there we have it — the date of Nebuchadnezzar's death: 37 years after 3405, that is the year 3442 after Creation.

But, that was a date known from *non-biblical* sources — it was 562 BC.

Here then was the linkage Ussher needed — two timelines intersecting at the one event. The year 3442 after Creation corresponded to 562 BC. That is, Creation took place 3442 years before 562 BC.

The conclusion — Creation occurred in 4004 BC!

Ussher had succeeded in determining how long ago it had been since God created the world. Quite an achievement. What's more — the date couldn't be more perfect. For it had been known for a long time before Ussher that Jesus had not been born at the intersection of the BC/AD axes. Really? How could that be? Blame Dionysius

Exiguus, the sixth century AD monk. He was the one who introduced the BC/AD system in the first place. But he made a mistake in figuring out how many years Jesus was born after the foundation of Rome. Later evidence and studies showed that the key figure, Herod the Great, King of Judea, had died in 4 BC. But according to Matthew 2:1, Jesus was born while Herod was still alive. So clearly Jesus' date of birth had to be on or before 4 BC! Usually it's taken as 4 BC.

So hallelujah! Creation in 4004 BC. Exactly 3000 years later, the building of a special Holy Temple for the worship of the Creator. Exactly 4000 years for the birth of the personification of that Holy Temple. Surely confirmation! The Archbishop must have felt in high heaven.

Perhaps that's what propelled Ussher to dig even deeper and try to find out not only the year of Creation, but which day of the year it was — and the time! By closely following what he saw as astronomical clues (eclipses, etc.), Ussher pinned down Creation to have taken place at "the entrance of the night preceding the twenty-third day of October" in the year 4004 BC. (All dates being according to the Julian calendar which was still in use in England in Ussher's 1650's; England — and its American colonies — did not adopt the modern Gregorian calendar till 1752.)

Ussher had planned to extend his chronology of world history. But, alas, that was not to be. Time ran out for him. In January 1656, sensing that the end was near, he ended his work for the day with a single word in large letters — "Resignation". He died two months later. Tracing world history was left to others.

Concluding remarks

Figure 4 shows a brief summary of Ussher's dates.

The left-hand timeline gives Ussher's chronology. From Creation onward. As he deduced from the Hebrew *Masoretic text* (mostly).

The timeline in the middle gives the corresponding dates in the BC/AD system. The crucial link between the two that Ussher used

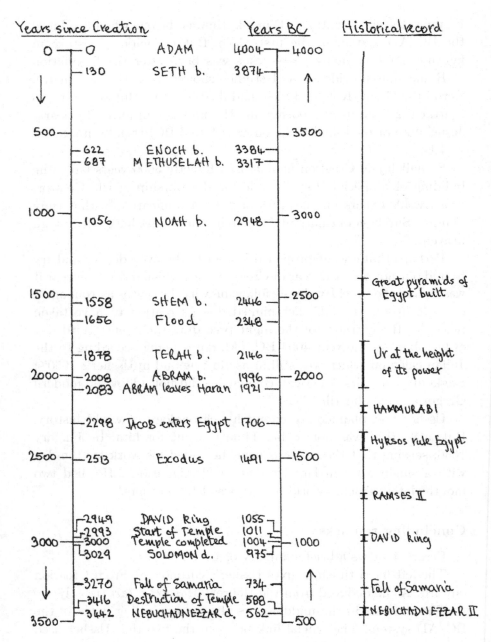

Figure 4:   Timelines from Creation to Nebuchadnezzar II.

was the year of the death of the Babylonian king Nebuchadnezzar II — 3442 in Ussher's chronology, 562 BC according to archaeology and historical records.

So voilà! Ussher could go back and find the BC/AD dates for his chronology. Those are the ones that appeared in the King James Version of the Bible.

However, it's certainly not true that the date 4004 BC for Creation on Genesis 1, page 1, was determined solely on the basis of the Bible. Ussher's use of non-Biblical sources unfortunately seems to have been forgotten. Sadly, it has generated an immense amount of hostility in the years that followed.

On the right-hand side of Figure 4, I have indicated time-segments for various well-known events as determined by modern scholars. What's amazing here is that, at least as far back as Abram (about 2000 BC), there's a remarkable overlap between the biblical and historical dates. For example, according to modern historians, David's reign was from 1010 BC to 970 BC. Ussher's dates were 1055 BC to 1015 BC. Pretty close. Or take the fall of the city of Samaria — 720 BC according to modern historians, 734 BC according to Ussher. And likewise for others. Impressive!

But the most important date of all was 4004 BC. Creation. And ever since that date appeared on the very first page of the Bible, many, many people have taken the age of the Earth to be about 6000 years.

And millions of people still do.

Appendix 1: The coming of the King James Bible

Without a doubt, the version of the Bible authorized by King James of England and Scotland in 1611 has been by far the most popular of all Bibles written in English. It was the culmination of the efforts of many scholars — often at great sacrifice — over the centuries. But two men stand out above all the others — the trail-blazers: John Wycliffe and William Tyndale.

John Wycliffe (1331–1384)

We take it for granted nowadays — having a Bible in our own language. Plenty to choose from. It's available in more than five hundred different languages. But it wasn't always that way. In John Wycliffe's day, in the 14th century, there was just one Bible — the *Vulgate*. And that was in Latin. You don't read Latin? You're not alone. Few ordinary people do nowadays. Even fewer back in the 14th century. Which meant that ordinary folk had no way of knowing for themselves what the Bible actually said. Just one of the things that incensed John Wycliffe.

Wycliffe was radical. He saw things a different way. No need for a Pope (never mentioned in the Bible); no need for that vast pyramid of powers within the Church; the monasteries with their clerics and extensive estates of great wealth should be dissolved (Henry VIII would have loved him); purgatory (a holding pen for unredeemed souls after death) didn't exist; no need for those church-enriching indulgences to release supposedly trapped souls and help them reach their final abode in heaven. And more. Not what the Church wanted to hear. Especially from a young upstart in a sleepy Oxford college with its tranquil lawns beside the gently-flowing Cherwell river.

The Bible was all-important to Wycliffe. To him, the Bible was the sole authority for all things religious. And surely *everyone* — not just Church officialdom — should have access to it. To guide their own lives. Which meant — the Bible had to be in English. Not Latin, the language of the *Vulgate*, the official Bible of the Church. But an English Bible didn't exist. At least not yet. True to his convictions, Wycliffe (with some colleagues) began the task of producing such a

Bible — an English translation of the *Vulgate*. The completed version (in manuscript form) appeared in 1382. The Pope was furious, the Church incensed. With the inevitable consequences — Wycliffe and his followers were condemned, his English Bible banned (and whenever possible, burned).

Wycliffe died not long afterward, in 1384. But the Church hadn't finished with him yet. It sensed its authority was being undermined. Threatened. So, over time, became more repressive. A consequence of which was that in 1415, it decided to denounce the long-dead Wycliffe as a heretic. Then in 1428, exhumed his bones, had them crushed and burned. The few remaining ashes tossed into a nearby river. To be dispersed. To leave no trace of his body on Earth and to banish his soul to oblivion.

But in spite of the Church's efforts, that wasn't the end of Wycliffe. What he had done was a beginning, the thin edge of the wedge. Others would eventually come forward to follow in his path.

And what about Wycliffe's Bible itself? Only moderate success. Many copies were made — secretly, all handwritten. Bulky. Then distributed. Often to end up in flames. The translation itself — very academic, stiff, often a stilted English too close to the original Latin. Further progress had to wait. A long time. Not until after the printing press had been invented (Johannes Gutenberg, around 1450), which offered the possibility of mass production. At a reasonable cost. It also needed a new firebrand to take up the torch of liberating the Bible from its Latin shackles and to breathe into it the fresh air of plain English. That genius was William Tyndale.

William Tyndale (1494–1536)

Tyndale was no ordinary translator. He was much, much more. A man with the golden touch when it came to words. So many of his succinct, powerful, resonant phrases have become part and parcel of the English language itself. Who doesn't know these gems?

> *the powers that be; a law unto themselves; a sign of the times; eat, drink and be merry; seek and ye shall find; fight the good fight; and in the twinkling of an eye;*

all from the New Testament, and

> *let there be light; let my people go; parting of the way; land of the living;*
> *pour out your heart; the apple of his eye; flesh pot;*

from the Old Testament. All vintage Tyndale. Short, powerful. Mostly Anglo-Saxon word-origin. Not Latin.

Like Wycliffe, Tyndale's inspirational dream was to make the Bible accessible to all, even (as he said) to the lowliest "boy that driveth the plow". But translating the Latin Bible into any vernacular language was still forbidden, even in the early 16th century. Extremely hazardous for any perpetrator. With Reformation in the air, the Church's stance against anything that challenged its power had become even harsher — not just excommunication and being declared a heretic. Penalties now included being tortured and burned alive at the stake. The soul damned forever.

England at this time was a great bastion of Catholicism. King Henry VIII himself was the leader, the Defender of the Faith — a title conferred on him in 1521 by Pope Leo X after the publication of Henry's book in which he had vehemently denounced Martin Luther's Protestant doctrines. England wasn't safe for Tyndale. So in 1524, he secretly moved to the more hospitable Luther – influenced Northern Europe. It was there he was able to work on his dream.

Two years later, in 1526, Tyndale's translation of the New Testament appeared. (Notably, it was based on the 1516 Greek translation by the Dutch scholar Erasmus, not on the *Vulgate*.) His translation of the Pentateuch (the first five books of the Old Testament) followed four years later in 1530. Of crucial importance was the fact that Tyndale's translation was based not on the Old Testament of the *Vulgate* but on the Hebrew *Masoretic Text* which he considered much more reliable.

But even Northern Europe wasn't safe. The "powers – that – be" had their spies. Agents not only of Henry but also of the Pope and the Holy Roman Emperor. On the lookout for this elusive heretic. Tyndale had to keep on the move. Which of course couldn't last forever. In 1535 came betrayal and arrest at a "safe" house near Antwerp in Belgium. A brutal eighteen months in prison followed.

Found guilty of course. Tied to a stake. Garroted by chain. Then burned — still alive. Horrendous. All for God's sake. A brilliant brave man. His last words — "Oh Lord, ope the King of England's eies".

Out of tragedy came triumph. Thousands of copies of Tyndale's pocket-sized English Bible were now pouring off the printing presses in Northern Europe and being smuggled into England. At last, ordinary folks could hear (and some read) the Holy Words in a language they could understand. Tyndale's dream had come true.

Winds of change

But even in England itself, times were a-changing. Led on by that Defender of the Faith himself, King Henry. As time went on, Henry wasn't quite so enthusiastic about defending the Faith he had written about earlier. Especially after Pope Clement VII refused to annul his marriage to Queen Catherine of Aragon. (Ironically, in his book, Henry had not only attacked Luther but also had vehemently defended the sacrament of holy matrimony *and* the complete supremacy of the Pope!)

It all came to a head in 1533. That was the year when Henry publicly married Ann Boleyn (they had been married in secret at the end of the previous year); his appointed Archbishop of Canterbury, Thomas Cranmer, having declared Henry's marriage to Catherine null and void, his marriage to Ann valid. Excommunication of Henry and Cranmer immediately followed. On the positive side however, Parliament quickly confirmed Henry as the supreme head of a new Church of England, and reinstated his grand title of Defender of the Faith. (British coins still have the initials FD or FID DEF stamped on them.) Of course, what was meant by "the Faith" had changed somewhat. 1533 was definitely an eventful year for Henry.

But let's not forget the Bible in all of this. Henry had banned Tyndale's English Bible in 1530. But he was a man who could change his mind (all too easily when it was for his own benefit). Incredibly, before long, Henry was commissioning a new Bible of his own. Not in Latin, but in English. It appeared in 1539, only three short years after Tyndale had been brutally put to death! And apart from some judicious word changes, it had incorporated much of Tyndale's banned Bible. Plagiarism was no problem for Henry.

Henry's Bible is known as the Great Bible. "Great" — not because it was some new superior translation, but because of its size! A copy was to be placed in each church of the land (at the church's own expense), at some place of convenience for the worthy parishioners. Apparently however, these same parishioners were not to be trusted — the Great Bible was to be chained in place! Hence, it's other name — the Chained Bible. It was the first authorized version of the Bible in English.

With the sluicegates open, other English versions of the Bible began to appear. Among them was the Geneva Bible, published 1557–1560. It was prepared by a group of scholars who had fled England during the reign of Henry's Catholic daughter Mary. They had settled in John Calvin's Geneva. And the basis of the Geneva Bible? Tyndale again. For the next fifty years, this Bible (also pocket-sized) was the Bible of choice for ordinary folks (including the pilgrims onboard the *Mayflower* in 1620, making their way across the Atlantic to America), and also for some of the not-so-ordinary (William Shakespeare, John Knox, and the author of *The Pilgrim's Progress*, John Bunyan, to name a few.)

But the popular Geneva Bible had too radical of a slant for the now established Church in England. So the next monarch, Elizabeth I, commissioned yet another version of the Bible. The outcome — the Bishops' Bible, so-called because of the large number of bishops on the revision committee. By most accounts, not a very inspiring version, many of Tyndale's simple, powerful Anglo-Saxon words and phrases replaced by stuffy old Latinized vocabulary. It was dutifully read by the priests on Sundays, but not by the people at home.

So what's going on here? Why so many versions of the Bible? Why not have just one correct translation and have done with it?

Trouble is — translation is not as easy as it may seem (as I well remember from my own seven years of Latin torture during my school days). And those things called words? Unfortunately, it often happens that the same word can have different meanings for different people.

An example. The Greek word εκκλησια (ecclesia). Should it be translated as "church" or "congregation"? To some people back in the 16th century, the word "church" implied not just a group of believers, but much more — the whole institutional organization with its hierarchical structure of bishops and priests. The word "congregation" in contrast suggests just a plain gathering of people with none of the implied power structure.

Likewise, for other pairs — bishop or overseer? Priest or elder? Charity (implying good works) or love (unconditional)? Confess or acknowledge? Do penance or repent? And so on. Given the potency of words, it's easy to understand why certain choices were made in some translations (such as the sovereign-initiated Great Bible and the Bishops' Bible), quite different choices in other translations (such as Tyndale's Bible and the Geneva Bible).

Which meant continual friction between the official church in England (strong on tradition) and the reformist Puritans (by faith alone).

The King James Bible

Quite a conundrum for James Stuart when he became King of England in 1603 following Elizabeth's death. He had already been King of Scotland for thirty-six years. As an infant, he had been baptized a Catholic like his mother Mary, Queen of Scots. But as a boy, he had been raised by a string of strict Presbyterians. Now he found himself Supreme Head of a Church of England with problems. Lots of internal squabbling. Which way should he go?

Just like politicians of today are wont to do, James called a meeting. There, the bishops and puritans could air their grievances and debate their differences. Let off steam. It was at this meeting (in 1604) that the suggestion was made to have a new, authoritative translation of the Bible. James liked the idea.

On the surface, it looked as if this would be a way of settling the differences. Make some compromises in the language. Both sides could accept the result. Trouble was — the canny Scot had a mind of his own. The English system appealed to him — simultaneously at the top of the political pyramid and also the religious one. Likewise, the Tudor belief in the divine right of kings — appointed by the Grace of God, subject to no-one, with absolute authority in *all* political and spiritual matters! James was no fool.

So while appearing to be open-minded, James stacked the deck to ensure a favorable (for him!) outcome. There were to be six "companies" of translators in all, two each from the religious powerhouses of Westminster, Oxford, and Cambridge. Each company would have about ten members, responsible for a portion of the Bible. The stated idea was not to make a new translation, but to "make a good one better". And which "good one" were they to start from? The Bishops' Bible. (We can see where this is going!) Moreover, ecclesiastical words such as "church", "bishop", "priest", etc. were to be kept. The idea of kingship was to be stressed throughout. "Congregation", "elder", "minister" weren't to be considered. Cross-references to other verses in the Bible were allowed, but definitely no explanatory notes (James considered some of the marginal comments in the Geneva Bible in particular anti-royal; his solution — no comments at all). And after each company had done its work, two smaller committees would review all the translations. Then, as the final adjudicator, the pro-royalist Archbishop of Canterbury himself, Richard Bancroft.

The outcome? The 1611 King James Version of the Bible, authorized by the king himself. Incredibly, the foundation again was the powerful language of that heretic Tyndale (though of course his name was never mentioned). It has been estimated that more than 80 percent of the New Testament and more than 70 percent of the Old Testament is Tyndale. Most of his marvelous words and phrases, some amended to sound even better read aloud, were there.

But there were a few changes, with choice of words more to James' liking. Which still made the Puritans bristle. I see, for example, that 1 Timothy 3:1 in my King James Version reads, "If a man desireth the office of a bishop, ..." James had won out.

In spite of these changes, however, the King James Version over time became the accepted version, much beloved by millions in the English-speaking Protestant world in subsequent centuries. (The Catholic Church itself produced an English version of its Latin *Vulgate* in the early 17th century — so much for its old heresy laws.)

In some quarters, the King James Version is of such high standing that, in spite of the fact that there were translations before the King James Version and many afterward, some people still think it is the only authentic Bible and that it has *always* been there. I'm reminded of the story about the old lady who accosted the new preacher at her church. It happened after Sunday morning service in which he had had the audacity to read the Lesson from a modern Bible. "Young man" she said, pointing an accusatory finger in his face, "If the King James was good enough for Jesus, it's good enough for us"!

Appendix 2: Archbishop James Ussher (1581–1656)

James Ussher was a bright Irish lad. Born into a well-to-do family in Dublin in 1581. Quickly rose through the political and ecclesiastical ranks to become Anglican Archbishop of Armagh and Primate of all Ireland in 1625. At the top, apart that is from the Archbishop of Canterbury himself back in England. And there's the rub. The two Archbishops didn't see eye to eye — on hardly anything. Neither with the running of the Church nor the politics of Ireland. Ussher's stern Calvinistic leanings were probably not quite the balm of Gilead that was needed, given the turbulent times of the day. (When are there not "troubles" in Ireland?) And so it was that, during one of his periodic visits to England, Ussher decided to stay there. For good. Never to return to his native land.

But if Ussher was expecting a more peaceful time in England, he was in for a big surprise. For it wasn't long before hostilities broke out there as well. This time between Charles I (the Catholic king, of divine-right-of-kings fame, subsequently beheaded) and his Cavaliers on the one hand, the Puritanical Oliver Cromwell and his Parliamentarian Roundheads on the other. Ussher managed to survive — but only by the skin of his teeth. Eventually taken care of by a good friend, the Dowager Duchess of Peterborough. At long last, he was able to settle in and devote all his time and energies to completing his beloved scholarly pursuits.

And the outcome? His *Annales*, a massive two thousand-page tome (in Latin) in two parts, the first part published in 1650, the second part four years later. A detailed history of the Western World, from the very day of Creation up to the destruction of the city of Jerusalem in AD 70 by the Roman general (soon to become Emperor) Titus. A monumental *tour-de-force*. It included his famous biblical chronology.

Over the years, Ussher had consulted a vast number of ancient books and manuscripts. Of particular importance, as far as his biblical chronology was concerned, were two versions of the Old Testament. They were based on different Hebrew traditions. The much older of the two — the *Septuagint* — was a translation by Jewish scholars in the sixth century BC of very old Hebrew documents (unfortunately now lost). Into Greek. For the benefit of the Greek-speaking, non-Hebrew-speaking, Jewish community in Alexandria, Egypt. The other was the Hebrew *Masoretic Bible*, prepared much later by Jewish scholars in the sixth to tenth centuries AD.

Of these two, Ussher relied primarily (but not exclusively) on the *Masoretic Bible* which he considered the more reliable. (The very same text that William Tyndale had used for his translation more than a hundred years earlier, much of which was incorporated into the Authorized King James Bible in 1611.)

But Ussher didn't finish there. For there were other sources that he felt might provide additional insights. Trouble was — they were written not in Hebrew, Greek, or Latin, but in other ancient languages such as Samarian, Aramaic, Chaldean, and Old Persian. Only a minor hurdle for our polyglot. He learned all of them! So that he could read them for himself. Not relying on other people's translations. Ussher left no page unturned.

Even in his day, Ussher's incredible achievement was recognized. A history of the western world from Creation till the Roman destruction of Jerusalem in AD 70. In detail. But it was the biblical chronology part — which was only about 15% of the whole book — that stuck. Thanks to Thomas Guy. Incorporating Ussher's dates into the increasingly popular 1611 King James Version of the Bible. 4004 BC was the date of Creation. And many still believe that today. Fervently.

[Modern Bibles no longer include Ussher's dates — they were dropped by the publishers Oxford University Press and Cambridge University Press in the early 1900s.]

Bibliography

It's useful of course to have access to a King James Version of the Bible which contains Ussher's dates. The copy I have (an old family Bible) is:

The Holy Bible. 1901. New York: World Syndicate Company, Inc.

It claims to be an "unchanged" edition, though it omits the various preliminary sections (such as the dedication to King James).

The initial publication of the King James Bible in 1611 did not go smoothly, and errors persisted for many years. Exodus 20:14 in the 1631 edition, for example, read "Thou shalt commit adultery" (having omitted the crucial "not"); it became known as the *Wicked Bible.* It is not known how many people obeyed that version of the Seventh Commandment. The edition in 1653 corrected previous errors but mistakenly informed readers (1 Corinthians 6:9) that "the unrighteous shall inherit the kingdom of God". Even as late as 1801, there were errors still appearing — Jude 16 read "these are murderers, complainers, walking after their own lusts". The "murderers" should have been "murmurers"; that version got the nickname *Murderers' Bible.*

Nowadays, there is a vast plethora of Bibles, all of which have even vaster quantities of explanatory footnotes and comments (just what King James vetoed in the version he authorized!). The modern Bible I used most often was

The Study Bible 1989. (gen. ed. W.A. Meek). New York: Harper Collins Publ. This is based on the Revised Standard Version of the Bible.

The best article that I know of on Ussher's chronology is

Barr, J. 1985. *Why the World was Created in 4004 BC: Archbishop Ussher and Biblical Chronology.* Manchester: Bulletin of the John Rylands Library **67**, 575–608.

It helped me enormously.

I have also benefited from the following four books which tell the story behind the King James Version of the Bible:

Bobrick, B. 2001. *Wide as the Waters.* New York: Simon and Schuster.
Nicholson, A. 2003. *God's Secretaries.* New York: Harper-Collins Publishers, Inc.
Campbell, G. 2010. *Bible: The Story of King James Version 1611–2011.* New York: Oxford University Press.

Gorst, M. 2001. *Measuring Eternity*. New York: Random House, Inc.

A more recent book is

Bragg, M. 2017. *William Tyndale — A Very Brief History*. London: SPCK.

Interesting articles on the subject are

Reese, R.L., Everett, S.M. and Craun, E.D. 1981. The Chronology of Archbishop
    James Ussher. *Sky and Telescope* **62**, 404–405.
Brice, W.R. 1982. Bishop Ussher, John Lightfoot and the Age of Creation. *J. Geol.
    Educ.* **30**, 18–24.
Brush, S.G. 1982. Finding the Age of the Earth by Physics or by Faith. *J. Geol.
    Euc.* **30**, 34–58.

Since I first started to delve into the "biblical" age of the Earth, the Internet
has been invented. Their search engines provide the reader with a vast mountain
of information.

Chapter 2

James Hutton and Infinite Time

"...no vestige of a beginning, no prospect of an end..."

I well remember the day I "discovered" Siccar Point. I was driving along the coastal road from Edinburgh in Scotland toward the town of Berwick further south on the English border. It was early afternoon. I had been expecting to come across a big sign that would direct me toward one of the most celebrated geological sites in the world. But no, there was none. So I stopped and asked a local farmer. Go back, he said. Turn right just before you get to the bridge. You'll end up in a quarry. Park there. Then head over the hills to the coast. You'll find Siccar Point there.

All of which I did. The cows didn't seem to mind. Finally, I came to the top of a hill. Beneath me, the ground sloped steeply down toward the waves of the North Sea. I knew at once I had found Siccar Point. It was staggering.

James Hutton (1726–1797)

Two hundred or so years earlier, in 1788, three others had "discovered" Siccar Point. James Hutton and two of his friends. They had come by boat along the coastline. Looking for evidence that would support Hutton's new theory about the Earth.

At that time, most people believed that the Earth was only about 6000 years old. Based entirely on the Bible, it was said. The dates were there to prove it, at the top of each page. And all geological structures seen on the surface of the Earth were somehow the result of that single biblical event, the Flood.

Hutton had other ideas. He had expounded them in two lectures three years earlier at the Royal Society of Edinburgh. To him, the Earth was a huge dynamic system — not constant, but *constantly changing*. Land was continually being uplifted out of the oceans, even to form high mountains. Then erosion, by natural processes — wind, rain, snow, ice, sun, and waves. Which broke up the solid rock. Washed down the resulting detritus into the oceans. There, it gradually compactified. Back to hard rock again. Then uplift to re-form the dry land. A non-stop cycle of uplift and erosion. All a slow, very slow process. Which could be repeated over and over again. So what about the age of the Earth? Hutton concluded it had to be old, very old. More than 6000 years for sure. A lot, lot more.

Hutton had not always been a geologist. Born in Edinburgh in 1726 to wealthy parents, he attended the University there, studying Latin and Greek. After that, a brief apprenticeship as a lawyer, then as a physician's assistant. Next, he enrolled as a medical student — first in Edinburgh, then Paris, finally Leiden in Holland where he graduated in 1749.

But medicine didn't seem to be his calling either. (Not every young person knows what he/she wants to do in life!) He returned to London, then back to Edinburgh. There, he joined forces with an old friend from student days. To develop a new chemical business. Nothing exotic — just the manufacture of salammoniac. Used for cleaning metal surfaces. A financial winner.

Then came farming. For after his father had died, Hutton (being the only son) had inherited the family estate — which included two farms about 50 miles south of Edinburgh. Moved there in 1752. Introduced the latest innovations in agriculture (he had studied them during a long stay in England). It was while he was there on the farm that Hutton's interest in matters geological finally took root. Impressed with the large variety of minerals he found on his land. And the rocks he had collected on the various expeditions he had made all over Scotland and England. Hutton had changed. Had caught the geology bug.

Now independently wealthy from his inheritance and the profits from his chemical business, Hutton made one final move. Back to his hometown, Edinburgh. 1768. In his 40s.

At that time, the nation's capital had become a place of tremendous intellectual excitement. Nowadays referred to as the "Scottish Enlightenment". Among the famous who were living there at the time were the great moral philosopher David Hume (of *Treatise of Human Nature* fame), the economist Adam Smith (*Wealth of Nations*), and the scientist Joseph Black (discoverer of carbon dioxide). Plus visits from such distinguished people like Benjamin Franklin and James Watt. Quite a galaxy. Formed the Oyster Club. Hutton was a founding member. Met weekly in local taverns. To indulge in the delicacy of the day (hence, the name). And for discussions.

Finally in 1785. The big step. Going public. Hutton was now almost 60 years old. Two lectures to the newly formed Royal Society of Edinburgh. On his ideas about the Earth — uplift and erosion; a continuous cycle; going on forever. Forever? Well, to Hutton at any rate, forever. The Earth was really old. With no beginning in sight.

So who's right? Who's wrong? A young Earth, or an old one?

What Hutton really needed was a clincher. A geological formation that could be understood *only* in terms of his ideas. Not biblical ones. Which takes us back to Siccar Point. That's where Hutton hit the jackpot.

Siccar Point

It's quite a sight. Figure 1 illustrates what I saw that day from the hilltop overlooking Siccar Point. It was a warm, sunny day — quite different from the rain-drenched field trips I remember taking as a geology student many years before. At Siccar Point, there was no ambiguity in what I saw. And it was in technicolor!

For, emerging out of the North Sea coastline in striking *vertical* layers was a thick stratum of rocks, gray in color. The joins between its many layers etched out by the pounding waves. And sitting on top of this gray stratum were layers of a totally different kind of rock. Sandstone, reddish in color. Nearly *horizontal*. Wow!

Figure 1:   Schematic of rocks at Siccar Point.

How could all that have come about?

One possibility — the biblical Flood, caused by a deluge that lasted 40 days and 40 nights. Could this all have happened as a result of one brief Flood?

For Hutton, Siccar Point provided a stark confirmation of his ideas. For there was a simple explanation to all of it, in terms of natural processes:

1. The vertical gray rock (called "greywacke" nowadays) was originally horizontal, at the bottom of a deep ocean, made up of loose detritus from the surrounding mountains; brought there by erosion, rivers, and ocean currents.
2. A gradual compactification of this loose material, by weight and pressure, forming the sedimentary greywacke.
3. Upheaval; in this case, uplift plus a sideways squeeze causing the greywacke to buckle.
4. Erosion of this uplifted land, all the way down to sea level; producing a horizontal surface that cut across the folds of the greywacke.
5. Submersion to great depths.

6. Deposition of layers of reddish material washed down from the land.
7. Compactification. Into a reddish sedimentary stratum. Horizontal.
8. Another uplift.
9. More weathering and erosion, reducing the exposed rock to what it looks like today. A reddish, almost horizontal stratum sitting on top of a gray vertical stratum.

A long story. Ordinary nature at work. No miracles. Is this how it all happened?

Like any good detective, Hutton searched for corroborating evidence. Which he soon found. Not far from where he had landed. A section of the vertical greywacke with ripple marks! Conclusion — these layers that were now vertical must have once been horizontal, lapped by water.

And at the base of the reddish sandstone, a layer containing fragments of the greywacke. Confirming that the erosion of the (vertical) greywacke had occurred before the (horizontal) sandstone stratum was formed. For Hutton, the evidence was overwhelming.

And who knows how many of these erosion – compactification – uplift cycles this part of the Earth (or any other part for that matter) had gone through in earlier times? As John Playfair, one of the passengers, later wrote: "the mind seemed to grow giddy by looking so far into the abyss of time".

But wait. There's something missing in all of this. Which needs an explanation. Uplift. What causes that?

Hutton had a revolutionary new idea for that too. Heat was his answer. Heat energy stored within the Earth, influencing the surface. Really? Was there any evidence for this heat?

Hutton certainly thought so. For among the many different types of rocks he had picked up on his farms and various explorations, there were some that looked quite different. Not like the ubiquitous sedimentary rocks with their distinct layers of fragments and fossils. These other kinds of rocks didn't have any layers at all. Or fossils. Instead, had lots of little crystals. Granite is what it's called. Often found in veins that cut through layers of sedimentary rock. What were they, and how did they get there?

For Hutton, it meant that these granites had at some earlier time been *molten*. Hence, no fossils. And how did these granites get there? Pressure. That squeezed the molten rock into any weakness or fracture in a sedimentary stratum. Often forming veins. Thereafter, slow cooling and gradually solidifying. Allowing crystals to form, their size depending on how fast or how slow the cooling took place.

One important consequence — the intruded granites were *younger* than the surrounding sedimentary rock. Just the opposite of the then – accepted understanding of granite. Hutton's contemporaries believed that granite was an old type of rock, in fact the *oldest* kind of rock there was, existing before life itself. Hence, no fossils. Diametrically opposite points of view.

Salisbury Crags

And there was more. For Hutton discovered — at least from his perspective — an even better example of intruded rock right in his own backyard. In Edinburgh. For Edinburgh is built around two striking geological features — Castle Rock (on which its famous castle is situated) and further east — Arthur's Seat (no one seems to know who Arthur was). They rise sharply out of the green plain of East Lothian. Both are volcanic in origin.

Castle Rock is by far the more popular with tourists. You can see the Crown Jewels of Scotland in the castle there. As well as the room where King James VI of Scotland (later to become also James I of England) was born to Mary, Queen of Scots. That's the James of the 1611 Authorized Version of the Bible.

The bare steep sides around much of Castle Rock suggest its origin — it's what's called a volcanic plug. There are quite a few of them in Scotland, including one only a few miles east of where I was born and grew up. Loudoun Hill, it's called; a little over a thousand feet high. The remnant of an old volcano, a big chunk of once-molten rock that lodged in the volcano's throat. The view from the top is spectacular. A Roman fort at its base, to the south. To the east, sand quarries — the sand deposited there during the Ice Age. And two battlefields — one in 1307 when King Robert the Bruce of Scotland defeated an English army, the other in 1679 when the rebelling Covenanters won their religious freedom. Toward the north, Lochfield farm where Alexander Fleming (the discoverer of penicillin) was born; and Glaister farm where I worked as a lad one summer. Westward, the valley of the River Irvine, opening up into the fertile plain of Ayrshire. Beyond that, the Firth of Clyde shimmering in the distance. In the Firth itself sits the Isle of Arran where Hutton found so many other geological sites to support his ideas. South of Arran, the small conical island of Ailsa Craig sticks up out of the water — another volcanic plug whose granite was quarried for centuries to make the world's curling stones.

Not a view you're likely to forget!

The view from Arthur's Seat in Edinburgh is likewise spectacular. Holyrood Palace at its base. And the new buildings of the Scottish Parliament. To the north, the wide Firth of Forth and the spectacular Forth Bridge. And Old Town, in the direction of Castle Rock. That's where, in Greyfriars Kirk cemetery, both James Hutton and his good friend Joseph Black lie buried.

For James Hutton, Arthur's Seat provided another of his special sites. The summit itself and nearby ridges are remnants of an old volcanic system. But further downhill, it's different — layers and layers of what is obviously sedimentary rock. To Hutton, it meant that at some time in the past, the whole volcanic system had been submerged for a long time, subsequently uplifted and eroded. What specially caught Hutton's attention however was something else. A thick dark layer of rock jutting out markedly from the hillside (Figure 2).

This strikingly different feature is known locally as the Salisbury Crags. No fossils. Fine-grained. Not sedimentary like the surrounding rocks. How did it get there?

Hutton had a simple explanation. The rock that forms the Salisbury Crags had once been molten. Under huge pressure, it had been squeezed into the sedimentary layers, along a line of weakness, splitting it apart. Followed by solidification. Then erosion, which exposed

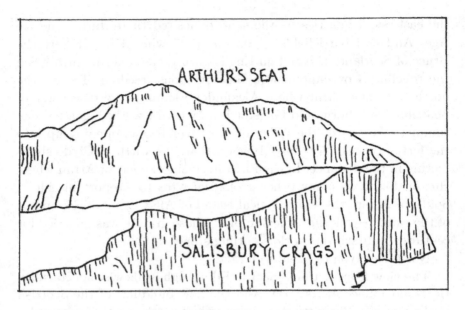

Figure 2:   Schematic of Arthur's Seat with the Salisbury Cargs.

the intruder. Clearly the injected rock was *younger* than the surrounding material, not older just because it contained no fossils.

Far-fetched? Not to Hutton. Especially when he again found corroborating evidence. At the interface between the sedimentary layers
and — as he thought — the intruded rock, Hutton found small pieces
of the sedimentary rock embedded *within* the dark rock. How could
these fragments have got there if the dark layer had not come *after*
the sedimentary layers and been molten?

A fascinating explanation for a puzzling geological feature. But,
like the thinner veins of granite he had previously identified, it
required a lot of pressure to squeeze that molten rock into place.
Where did such huge pressures come from?

Hutton again had an answer — thanks to insight from his Oyster
Club colleague Joseph Black. In his younger days, Black had made
careful studies of heat. In particular, he showed that heat could be
stored or released, even when there was no change of temperature —
he referred to it as "latent heat" (it's still called that today). Hutton

was sure there had to be a big heat reservoir within the Earth. And this was surely the source of the power that drove volcanic activity and intrusions of molten rock. And over time, this heat could also be the source of the power that was able to raise vast regions of solid rock to heights of thousands of feet.

Clearly a man of great insight.

Passing the torch

Needless to say, few were convinced. Most stuck to the biblical story. A young Earth. Which all happened in a tick of geological time. Hutton responded with a two-volume tome entitled *Theory of the Earth*, published in 1795. Judged by all to be deadly dull. Unlike his lectures apparently. Perhaps, it was because Hutton was seriously ill. He died three years later.

It took someone else to carry the torch for Hutton's ideas — now called "uniformitarianism". A big awkward word. But it stuck. The man who made it a key idea was fellow Scot Charles Lyell (1797–1875). Famous for his *Principles of Geology*, first published as three volumes in 1830–1833. It incorporated and substantiated Hutton's ideas. Deep time. Exciting to read. Volume 1 turned out to be particularly significant — just before setting off on his momentous round-the-world voyage on the HMS *Beagle* in 1831, Captain Robert Fitzroy gave a copy of it to the young naturalist who was to be his paying companion on the voyage — Charles Darwin.

Bibliography

In this chapter, I have made use of three biographies of James Hutton:

Repcheck, J. 2003. *The Man Who Found Time*. Cambridge, MA: Perseus Publishing.

McIntyre, D.B. and McKirdy, A. 1997. *James Hutton: The Founder of Modern Geology*. Edinburgh: National Museums of Scotland Publishing Ltd.

Dalziel, I.W.D. 1999. *Vestiges of a Beginning and the Prospect of an End*. In Craig, G.Y. and Hull, J.H. (eds) *James Hutton — Present and Future*. London: The Geological Society.

There is of course nothing that can replace visiting important geological sites. Which I have been fortunate to be able to do — by walking, bus, car, and motorcycle — in my native Scotland. Amassing a small collection of rocks and fossils. Which is a lot smaller now than it used to be — when my parents moved house after I left for a postdoctoral physics position in the States, my mother (bless her soul) threw my collection out as "junk". Fortunately, I was able to retrieve a few pieces — the ones she had used to decorate the pathway at the front of her house. They included my chunk of shiny black galena from a disused lead mine in southwest Scotland, and a sliver of Torridonian sandstone from Beinn Eighe in northwest Scotland. The latter — a billion years old. Which I still have.

Chapter 3

Charles Darwin and the Weald

"...the denudation of the Weald must have required 306, 662, 400 years..."

There's only one calculation in the whole of Charles Darwin's famous *On the Origin of Species by Means of Natural Selection*. The book was first published in 1859. Darwin was 50 years old at the time — no youngster. It was an immediate best-seller. The entire first edition (1250 copies) was sold out on the very first day. The calculation itself is there in Chapter IX. Within a few weeks, Darwin was regretting his decision to include it.

The Origin of the *Origin*

Part of the problem was of course that the *Origin* had been a rush job. Darwin was in panic mode when he wrote it. He was scared he would lose the credit he felt was his due for the idea of evolution by natural selection. His friends advised him to publish quickly. Which he did.

So how did this all come about?

The young Darwin

In his youth, much had been expected of young Charles. He came from a distinguished family. His father Robert — a wealthy physician and financier. His father's father Erasmus — likewise a physician, also a natural philosopher and poet; widely regarded as a highly gifted individual. His maternal grandfather — Josiah Wedgewood. Famous for his potteries in the Midlands. They produced some of the world's finest china. Still highly prized. As we might say today — Darwin inherited good genes (though of course genes weren't discovered till the 20th century). So, following the family tradition and with some strong "encouragement" from his father, Charles enrolled as a medical student in 1825 at the University of Edinburgh in Scotland (at that time, considered to be the best medical school in Britain). Not a good idea. Darwin couldn't stand pain or even the sight of blood! Instead, he spent his time learning about all sorts of marine creatures and the art of stuffing animals. Much to the annoyance of his father. Who switched him after the second year to the University of Cambridge. Closer to home. Where he could keep an eye on him. And for a much more worthy cause — preparation to become a man of the cloth in the Anglican Church. Another bad idea. Darwin was more interested in beetle collecting and botany.

Then good fortune — at least from young Darwin's point of view. Through connections and family support (read "cash"), he was offered and accepted (initially against his father's will) the unpaid position of naturalist and travelling companion to Captain Robert FitzRoy on the HMS *Beagle*. A two-year voyage to survey and chart the coast of South America. To mark the occasion, FitzRoy presented Darwin with a copy of the first volume of Lyell's historic book *Principles of Geology*. It was to have a huge influence on the young naturalist.

And so it was, on the 27th day of December, two days after Christmas, in 1831, the HMS *Beagle* with Captain FitzRoy, Darwin, and a crew of seventy-three pulled away from Plymouth Sound. The same place from which the *Mayflower* had sailed in 1620, carrying the Pilgrim Fathers to a New England. And before that, in 1588, the *Revenge* under the command of Sir Francis Drake, leading the English fleet out to do battle with the mighty Spanish Armada. The *Beagle*'s departure would likewise presage a historic event.

The eventual itinerary was mind-boggling. First, the hot, volcanic Cape Verde Islands. Then across the Atlantic to Brazil. After that, south, along the east coast of South America. To Tierra del Fuego. Then north along Chile. Up the west coast, to another set of volcanic islands, the Galapagos Archipelago. Not finished yet. Across the Pacific to Tahiti. New Zealand, Australia, Cape of Good Hope. Back in the Atlantic. St. Helena (of Napoleon fame). The Azores. Then finally, on 2 October 1836, home. Quite a bit longer than expected — the voyage had lasted not two, but five years!

Darwin wasn't much of a sailor. Constantly seasick. But on land, it was a different story. An intrepid hunter-gatherer. Exploring on foot or on horseback. The coastlines. Major rivers. Forests, pampas. Mountains and volcanoes. Including three long trips into the Andes. Always observing. Writing it all down. Collecting, collecting. Whenever possible, shipping his specimens back to his mentor in Cambridge, Botany Professor James Henslow.

And what a massive collection it was! Birds (dried, stuffed), animals (skinned or bottled), snakes, lizards, fish, insects (especially beetles!). Plants, seeds. Shells, bones, skulls, teeth, tusks. A vast number of fossils. Some incredibly huge. All "useless junk", according to Captain FitzRoy. How wrong he was. Much of it was unknown, never seen before. Darwin was famous long before he set foot in England again.

But Darwin was more than a collector. He may have been an amateur before his voyage. By the time he returned, he was an experienced and seasoned scientist. With lots of questions. And ideas.

One thing for sure — during the voyage, Darwin had become totally convinced that Hutton (via Lyell) was right. Those layers of shells, for example, exposed on cliff faces high up on mountains — how did they get there? Surely they must have been formed at the bottom of the ocean, and subsequently uplifted. Which must have taken a long, very long time.

Darwin had even seen uplift in action. He witnessed earthquakes on several occasions. But the most stunning of all — arriving on the *Beagle*, on 4 March 1835, at Talcahuano Harbor in Chile. Or what was left of it. There had been a severe earthquake a few days earlier. Which had totally devastated the town. Hardly a single building left standing. And much of what had once been the harbor was now filled with rock that only days before had been the seabed. Now that seabed was several feet *above* sea level! Indeed, from all of his observations, Darwin concluded that the whole Andes mountain range that runs along the west coast of South America was rising. Extraordinary — but confirmed later in the 20th century ... to be understood in terms of the dynamics of plate tectonics.

And what about all that "useless junk"? It created a sensation back home. Fossils of large numbers of creatures. Many with no relation to anything then known on Earth. What were they? How long ago did they live? What had caused their demise?

And not just the fossils. So many of the plants and animals — never seen before. Were they in any way related to anything already known? Connections?

Questions, questions. How did it all fit together? It took Darwin another two years before the mist began to clear.

Beginnings of a theory

Even during the voyage, Darwin had noticed some peculiar facts. Take his *mockingbirds*, for example. As the *Beagle* went from island to island in the Galapagos, Darwin became aware that the mockingbirds on each island were noticeably different from one another. Why was that? He had also been told that it was possible to tell which island a *tortoise* came from just by the design on its shell. (The name "Galapagos" comes from the Spanish word for tortoise.) Intriguing. What did it all mean?

Then back home in early 1837, another eye-opener. From his ornithology friend John Gould who had taken on the task of identifying and classifying Darwin's bird collection. (Other experts worked on mammalia, fish, reptiles, and so on.) Those small birds that Darwin had collected on the Galapagos — they were all *finches*, a new group comprising of no fewer than twelve previously unknown species. What a discovery! But wait — there was a problem. The expert had goofed. Not Gould, but Darwin. For Darwin had inadvertently failed to label the particular island where he had captured each bird. A bad mistake. But was he lucky! Fortunately for him, others aboard the *Beagle* (including Captain FitzRoy) had also collected some specimens *and* labelled them by island. So with these other specimens, Darwin was able to reconstruct which island each finch had come from. The result — different species came from different islands. Just like the mockingbirds. And the tortoises. How come?

One way of explaining this conundrum was the then-popular idea of "centers of creation". In this theory, each island was to be considered a separate center of creation, with its own unique species of finch, say. Which stayed exactly the same forever. Never to change. The same now as always.

But Darwin had a problem with this. It occurred to him that, even though the finches from the various islands were all distinct from one another (such as by the sizes and shapes of their beaks), all of them had much in common with the finches he had seen on the South American mainland (about 560 miles to the east). So, it seemed more likely to him that, at some point in the distant past, finches from the mainland had managed to reach the Galapagos, possibly carried there by a violent storm. Then, after this initial "colonization" as he called it, the finches lived and bred. Somehow, by 1835, they had evolved into new species. Island-specific. How could that have happened? A big puzzle.

Then eureka! October 1838. Two years after his return to England. Darwin stumbled on to a crucial clue. As he described it in his *Autobiography*, that was when he "happened to read for amusement" the 1798 essay on the *Principles of Population* by the English cleric and scholar Thomas Robert Malthus. Malthus was concerned with the relationship between human population growth and poverty. He noted that, if unchecked, population would grow much faster than the availability of food. Which would eventually lead to catastrophe — starvation, disease, poverty, and war. (Still true today.)

So how to avoid such human catastrophes? For Malthus, the solution was to impose restrictions on human behavior, by law if necessary. The population could thereby be maintained at a sustainable level, pre-empting a vicious struggle for survival. (Malthus's suggestion of humans voluntarily limiting family size is popular today; some of his other ideas however were far from being charitable — he blamed the poor for their own miseries and opposed any kind of relief, sadly an opinion that can be heard all too often around the world even today.)

It was in this thesis of Malthus that Darwin found the kernel of the idea that was to become the evolution of species by natural selection. And from this moment onward, Darwin began to lead a double life. Most of his time was devoted to a vast number of research projects — writing about his long voyage on the *Beagle*, analyzing his huge collections, communicating with many other naturalists around the world (including a Mr. Wallace in Borneo), writing articles and books on geology and coral reefs, as well as on his beloved barnacles (which he studied intensely for seven long years). His expertise became widely recognized. The science community showered him with medals and honors. In secret, however (only a few of his closest friends knew), Darwin was working on another project — trying to figure out evolution, if that is how different species originated. He kept quiet about it since his thoughts all clashed with the accepted doctrine of his day — the Earth was only 6000 years old and all creatures had been created already perfect, totally suited to the environments they were in. Immutable.

But Darwin was highly aware of the fact that characteristics in plants and animals *could* be changed, and *had* been changed — often — by man himself: horticulturalists — to make new colorful varieties, farmers — to improve crop yield, horse breeders — for faster thoroughbreds, pigeon fanciers — for more varied looks. (Darwin himself kept a large flock of pigeons for experiments of his own.) In all these cases, change *had* been brought about successfully, by selecting out desired characteristics. So how could changes take place *without* the intervention of man — possibly even to the extent of creating new species?

For Darwin, Malthus held the key — ever-growing population, shortage of food supplies, struggle for existence, winners, and losers. To explain evolution, all you had to do was to apply "the doctrine of Malthus ... with manifold force to the whole animal and vegetable kingdoms". How would that work? Three main elements:

1. **Variation**: When creatures have offspring, the parents and offspring have many characteristics in common; they are alike. But not identical (not even "identical" twins, which I can vouch for

personally!). There are small differences between the offspring, right from the start.

2. **Selection**: The offspring are born into a particular environment. Which may have a plentiful supply of food. Or a lack of food that threatens existence, thanks to droughts or floods, scorching heat or freezing cold. Or predators — animals, insects, diseases. You name it. So, which (if any) of the offspring are going to survive? Those that have *inborn* characteristics which can enable them to adapt better to the circumstances they find themselves in, and survive ("survival of the fittest"). Selection, not by man, but by Mother Nature herself ("natural selection").

3. **Heredity**: Those offspring which survive are the ones that will reproduce. Thereby passing on to the next generation the special characteristics that enabled them to survive. They in turn pass these same characteristics on to succeeding generations. "Descent with modification", as Darwin put it.

With this process, Darwin could see how to explain, for example, the formation of different species on the various islands of the Galapagos — because of the slightly different environments on the islands, finches, mockingbirds, etc. with slightly different characteristics would gradually evolve on each island. And after many generations, new species. Island-specific.

But how to tell the world? And when? Well, not yet anyway. Darwin did work up what he called a "sketch" of his ideas in 1842, which he expanded two years later. But not quite willing yet to stand by his beliefs and go public. At one stage, he did start to write his masterpiece — a "big book", as he called it, to tell all. Then more procrastination. Still some doubt. Maybe he should find additional evidence to make his arguments even more convincing ... and so it went on, for years.

Alfred Russel Wallace

Then it happened. 18 June 1858. Out of the blue. An essay entitled "On the Tendency of Varieties to Depart Indefinitely from the Original Type", plus a cover letter, from the naturalist Alfred Russel

Wallace. Still thousands of miles away in the Malay Archipelago. It's what Darwin had been dreading. He had been scooped.

Wallace was fourteen years younger than Darwin. As a young man, he had been inspired by Darwin's *Voyage of the Beagle* (1839) and had gone off (self-supported) with a friend to explore the great rainforests of the Amazon. Amassing a vast collection of insects, reptiles, birds, and small animals. Four years' worth (1848–1852). Then tragedy. On his way back to England, in the middle of the Atlantic, the ship he was travelling in caught fire. Sank. Luckily, Wallace survived, rescued. But not his collection, nor most of his notebooks. They all went down with the ship.

Discouraged. But not broken. Two years later (1854), Wallace set out again, this time to explore another distant part of the world — the far-off Malay Archipelago. It lasted eight years. Again he collected a huge treasure trove of specimens. Labelling them. Noting their geographic distribution. Ahead of his time, Wallace was also very interested in the indigenous peoples — studying their customs, their interactions, and their languages (Wallace himself learned to speak Malay as well as several of the tribal languages).

In February 1858, however, Wallace was experiencing severe health problems — suffering from yet another bout of malaria. On the small island of Ternate, in the Moluccas. One benefit — it gave him time to think. About "the possible mode of origin of new species". Again, it was Malthus's essay on uninhibited population growth (which Wallace had read a dozen years earlier) that provided the inspiration.

Lying on his bed, it occurred to Wallace that "these checks — disease, famine, accidents, wars etc. — are what keep down the population ... then there suddenly flashed upon me the idea of the survival of the fittest ... that in every generation, the inferior would inevitably be killed off and the superior would remain ... considering the amount of individual variation that my experience as a collector had shown me to exist ... I became convinced that I had at length found the long-sought-for law of nature that solved the problem of the origin of species". Wallace immediately wrote up his ideas, willing to go public. Confident he had found the truth. Unlike Darwin,

Wallace had the courage of his convictions. Sent off his essay to the expert back in England.

So, what was Darwin to do? He immediately contacted two of his closest friends. They were the ones he had earlier confided in. Their solution — they would take care of it. By making a presentation at the next meeting of the prestigious Linnean Society of London, reading two notes on natural selection: one that Darwin had written to them years before, the other — the one that Wallace had sent to Darwin. The priority of the discovery would be shared. That date — 1 July 1858 — can be described as the official date for the birth of the theory of evolution by natural selection. Neither of the co-discoverers was present: Darwin was grieving at the time (his youngest son had just died of scarlet fever); Wallace had never been informed (he didn't return from the East till 1862 — fortunately with all of his collections intact this time).

And what to do next? Darwin realized that he had to get that "big book" out. Quick. But maybe first an "abstract" of it. Which he did. That "abstract" is the famous *On the Origin of Species by Means of Natural Selection*. Published seventeen months later, on 25 November 1859. Done in a hurry. The "big book" never materialized.

The calculation

If Darwin's theory for the evolution of species was correct, it needed time. Lots of it. Which is why Darwin favored geology's long view of the Earth. His own geological observations in South America and elsewhere seemed to support this picture.

But could he put a number on it? Or at least part of it? Darwin thought he could. The evidence was there, right in front of him. Literally!

By that time (1858), Darwin had settled with his family (eventually to include 10 children — obviously he had ignored Malthus's advice about limiting family size) in Down House, about 20 miles south of London. In the bucolic Kent countryside. Away from all the pressures and intrusions of the big city. To the south of him were two prominent ridges, about 800 ft high. They're called the North and

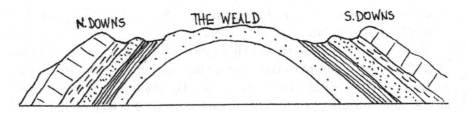

Figure 1: Stratigraphy of the North and South Downs.

South Downs ("downs" from an old Celtic word meaning "hill"). Both chalk. Between them, a shallow valley known as the Weald (after an old Anglo-Saxon word for "woodland").

Geologically, the Downs are the remnant of a large chalk dome, the top of which had been eroded away over time. Figure 1 depicts a cross-section of the area.

So what calculation did Darwin have in mind? It involved the Downs. How long, he wondered, would it take for that original dome of rock to be eroded away to its present state? He referred to it as "the denudation of the Weald". It would give him an estimate of at least one particular span of geological time.

Fortunately, it's an easy calculation. Spelled out in Chapter IX of the *Origin*. Darwin needed just three numbers.

First, the distance between the North and South Downs. It's about 22 miles.

Next, the thickness of the chalk stratum — about 1100 ft.

Lastly, the rate at which erosion takes place. That's a trickier number to get hold of. Darwin imagined a situation similar to what is happening to the famous chalk cliffs of Dover, a few miles further south on the English Channel. The waves there are slowly demolishing the beautiful white cliffs. Darwin's estimate was that "the sea would eat into cliffs 500 feet in height at a rate of one inch in a century".

Darwin realized this number was only approximate. He mentioned, for example, possible effects of different layers of the stratum having different hardnesses and of intermittent rock falls. Nonetheless, he concluded confidently that "for a cliff 500 feet in height, a denudation of one inch per century . . . would be an ample allowance".

Now for the calculation. Just two steps.

- If it takes a century for a 500-ft cliff to erode by 1 inch, how long would it take a 1100-feet thick layer to erode 1 inch? Certainly *more* time. Using direct proportions, the answer would be $(1100/500)$ centuries.
- If it takes $(1100/500)$ centuries for 1 inch to erode, how long would it take for 22 miles to erode? Changing these miles to feet, then inches, we get

$$(22 \times 5280 \times 12) \times (1100/500) \text{ centuries;}$$

that is, 306,662,400 years.

Which is the number Darwin gives in the *Origin*. About 300 million years.

Darwin must have been pleased with what he found. A long time. He had no hesitation including it in his book. And this was only part — "a mere trifle", as he put it — of geological time. The Earth itself must be very much older. Certainly plenty of time for the slow process of natural selection to take place and for species to evolve.

What a triumph!

But alas, no rest for the weary. Within a few weeks after publication, Darwin came under strong criticism, especially on this very calculation! His estimate for the rate of erosion, his critics screamed, was totally unjustified. It was ridiculous, for example, to use simple proportions. The erosion rate could have varied over time — certainly no reason to assume the rate in the past was the same as it is today. It might have been a thousand times faster, or slower! The 300 million years was totally unreliable.

The time calculation was actually the *last* place where Darwin had expected to be attacked. But as he reflected on it, he began to think his critics had a point. Maybe he hadn't given it quite enough attention. Been in too much of a hurry. So what happened? Darwin blinked. And by the third edition of his *Origin of Species* in 1861, Darwin had omitted the calculation altogether.

But it got worse. A famous professor up in Scotland was telling the world the Earth couldn't possibly be nearly as old as Darwin wanted. *Physics* said so. And who could argue against that? The man in question was William Thomson. Better known nowadays by his subsequent title of Lord Kelvin. Definitely a force to be reckoned with.

Bibliography

Darwin's story of his five years of exploration and discovery as a young man on the *HMS Beagle* is truly enthralling. It's a must read for anyone interested in Darwin:

Darwin, C.R. 1845. *The Voyage of the Beagle* 2nd ed. Reprinted by Santa Barbara, CA: The Natural Press.

As regards Darwin's *Origin of Species*, I was fortunate to receive a copy of a recent illustrated edition from my wife on our second wedding anniversary. Just what a non-biologist person like myself needed. I recommend it highly:

Darwin, C.R. 1859. *On the Origin of Species: By Means of Natural Selection, or the Preservation of Favoured Races in the Struggle for Life.* In Quammen, D. (ed); illustrated edition 2008. New York: Sterling Publishing Co., Inc.

Darwin's discussion of the Weald is in Chapter IX.

Besides Lyell's *Principles of Geology*, Darwin took with him on his voyage the seven-volume *Personal Narrative* by the great German explorer and naturalist Alexander von Humboldt. This described in detail Humboldt's five-year journey (1799–1804) through what is now Venezuela, Columbia, Ecuador, Peru, and Central Mexico. It was an inspiration for the young Darwin — "My admiration of his famous personal narrative (part of which I almost know by heart) determined me to travel in distant countries, and led me to volunteer as naturalist in her Majesty's ship *Beagle*".

Humboldt's fascinating life is described in

Wulf, A. 2015. *The Invention of Nature: The Adventures of Alexander von Humboldt, the Lost Hero of Science.* London: John Murray (Publishers).

Humboldt's great theme was the close interconnectedness of Nature, including mankind. Something that society has recently re-discovered ... von Humboldt was there first, long ago!

Wallace's story is described well in:

Winchester, S. 2008. *Krakatoa.* London: Penguin Books Ltd.

Chapter 3 describes both Wallace's discovery of natural selection, and what came to be called the Wallace line — an abrupt line of demarcation that separates two totally different biological regions, Asian fauna and flora to the west of the line, Australian to the east. A hundred years later, it was recognized as the dividing line between two huge tectonic plates that had shifted markedly in time.

Captain Robert FitzRoy wasn't much older than Charles Darwin when they both embarked on the historic voyage of the *Beagle* in 1831. Darwin was 22, FitzRoy only 26. He had been given command of the *Beagle* during its previous voyage when its captain had taken his own life.

And after the famous voyage, Captain FitzRoy just didn't disappear into obscurity. On the contrary, a year after his return, he was awarded a gold medal by the Royal Geographical Society and became a Member of Parliament. Then in 1843, he was appointed the Governor of New Zealand. There, in an argument between the indigenous Maoris and the land-grabbing colonists, FitzRoy sided with the Maoris. Which didn't make him popular in New Zealand. Even less with the powers – that – be in London who promptly recalled him ... permanently.

But FitzRoy's major legacy was still ahead! For after a devastating storm at sea in 1859 (the same year as the publication of Darwin's *Origin of Species*), FitzRoy began to develop a national system for what he called "forecasting the weather". The key component was the newly developed electric telegraph. Which provided him with essentially instant reporting. Each day, from fifteen land stations around the country, FitzRoy received local weather reports. From these, he made his forecasts. Soon they even became a regular feature in the national newspapers. Admittedly primitive and often wrong (not unheard of, even in these days of mega-computers). But it was the kernel of a great idea. The beginning of the now internationally famous British Metrological Office.

Sadly, in 1865, at the age of 59, plagued by financial worries (he had spent much of his personal fortune on various public projects) and attacks from the powerful nay-sayers of his forecasting, FitzRoy took his own life.

Fitzroy's story is told for example in:

Gribbin, J. and Gribbin, M. 2004. *The Remarkable Story of Darwin's Captain and the Invention of Weather Forecasting.* London: Yale University Press.

Chapter 4

Lord Kelvin and Thermodynamics

"...the consolidation of the Earth's globe cannot have taken place less than 20,000,000 years ago ... nor more than 400,000,000 years..."

It was early 1862. William Thomson — later Lord Kelvin — was frustrated. For two reasons. First, he was in constant pain. A year or so earlier, he had slipped on the ice while playing a game of curling. A bad leg fracture. Which was taking an inordinate amount of time to heal (it left him with a permanent limp). Second, he was being ignored. Shouldn't happen to someone of such national stature. His recent achievements — in both the academic and public arenas — had been stellar and widely recognized. He had devised an important new temperature scale (an *absolute* scale); discovered a basic law of Nature (the Second Law of Thermodynamics); found previously unknown connections between electricity and heat flow. Plus widespread acclaim for his key contributions to the successful laying of the first transatlantic telegraph cable in 1858. (Not his fault that the chief engineer ruined it shortly thereafter, zapping it with a devastating 2000 volts.) Kelvin was clearly a man of many talents. And he was still only in his 30s!

Trouble was — Kelvin disagreed totally with the geologists of the day and their favorite theory of uniformitarianism ("the past like the present"). And now the biologists (read Darwin and co.) were also demanding vast, stable stretches of time. To let their new-fangled idea of evolution-by-natural-selection work its magic.

According to Kelvin, it couldn't be right. Not possible. There could have been no vast eons of stability. *Change*, said Kelvin, is what happens over time, not stability; guided by "an overruling creative power". And how did Kelvin know all this — at least the science part? Physics. With its undeniable, universal laws.

Start with the Sun. A hot body. Cooling down. Its energy being radiated — "dissipated" — into space. *Irreversibly.* Consequences? The Sun must have been very much hotter in the past than it is at present. At a much higher temperature back then. Radiating off much more heat than now. Causing different climate conditions on Earth compared to what we have today. Storms more extreme. Floods and droughts more devastating. Earthquakes and volcanic eruptions immensely more violent. Obviously, a past *very different* from the present! The geologists were wrong, just plain wrong.

Kelvin was not just a man of words. He had, over the years, put a lot of work into understanding the Earth and the Sun. Published a series of articles and given many talks on the subject. Including his inaugural address at the University of Glasgow when he was appointed in 1846 — at the tender age of 22 — to the distinguished chair of Natural Philosophy. Its title — *De distrionibus caloris per terrae corpus.* (Yes, apparently he did give that address in Latin). And by 1854, he had even pinned down the source of solar energy — in-falling meteoric particles. All within the Earth's orbit, spiraling in toward the Sun. All other possibilities ruled out.

Not much of a response from the geologists. No evidence, they said, for Kelvin's cataclysmic events in the geological record. Paid no attention to him.

Now it was eight years later. 1862. Time for real action. Kelvin wasn't bashful. Always confident. The bottom line, the undeniable truth — the laws of physics could not be ignored. By anyone. Out of frustration, Kelvin produced two landmark papers: *On the Age of the Sun's Heat* and *On the Secular Cooling of the Earth.* Incorporated his latest ideas, with numbers. Both the Sun and the Earth, he concluded, were only about 100 million years old.

That finally got everyone's attention.

How did Kelvin figure it all out? Something I've wanted to know about for a long time. The details. Not just the history. How was he able to wave his wand and come up with these amazing numbers?

As you might expect, Kelvin's method was ultra-scientific. Educated guesses, calculations, consequences, and comparison. Even the briefest glance at his papers shows that the man was a veritable calculating machine. Relentless in his pursuit. A Sherlock Holmes, long before that infallible detective had ever been created. And Holmes' famous dictum — "When you have eliminated all that is impossible, then whatever remains — however improbable — must be the truth"? Kelvin already knew this. It was his mantra.

Let's start with the Sun. (Though if you're in a hurry to find out about Kelvin's treatment of planet Earth, you can skip this section.)

Measurement of the Sun's energy output

Kelvin lived at the right time. For in 1837, an experiment of fundamental importance had been done. By one, Claude Pouillet, of France. Pouillet had measured the rise in temperature of a known weight of water exposed to sunlight for a given length of time. From which you can figure out the amount of solar energy incident per second per unit area at the Earth's surface. The result, as quoted by Kelvin, was that the amount of solar energy incident per second per square foot at the *Earth's* surface is 0.06 thermal units centigrade.

What? Not familiar with "thermal units centigrade"? You've plenty of company. Few people nowadays are familiar with the old-fashioned system of units that Kelvin used. The international metric system wasn't universally accepted back in the mid-nineteenth century, though Kelvin himself became an ardent supporter and spokesman for the metric system. In 1902, for example, he made a personal presentation before the US House Committee on Coinage, Weights, and Measures to plead its cause. As we know, his appeal fell on deaf ears.

So, to help in our calculations later, I've made a list in Appendix 1 of various units we'll need, plus some formulae, numerical values, and conversion factors. One thermal unit centigrade is in fact the amount of heat necessary to raise the temperature of 1 lb of water by one degree Celsius ("centigrade" being the same as the more modern "Celsius").

Back to Pouillet. What's so important about his number? As Kelvin realized, it tells us something important about the *Sun*. For, as indicated in Figure 1, all the energy that radiates out through a large spherical surface whose radius R_E is the radius of the Earth's orbit (and on which the Earth lies), originally radiated out from the Sun's surface (radius $R_\odot$). Equating these two energies (see Appendix 2) yields the result that the energy radiated outward from the Sun's surface is 2784 thermal units centigrade per second per square foot.

One more step. Which amounts to converting *heat energy* into an equivalent amount of what Kelvin called *mechanical energy*. Such as energy associated with motion (kinetic energy, as Kelvin named it), or energy associated with a change of position under the influence of the force of gravity (called gravitational potential energy).

It was Kelvin's good friend James Prescott Joule who, in the 1840s, had successfully pinned down the quantitative relationship between these two different kinds of energy — and thereby laid the

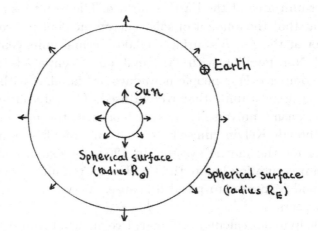

Figure 1: All the heat energy radiated outwards from the Sun's surface passes through a sphere whose radius is the radius of the Earth's orbit around the Sun.

foundation for a basic law of Nature, the First Law of Thermodynamics. It involved a whole series of impressive experiments — both
mechanical and electrical. Such as the one where he measured the
tiny rise in temperature of a known volume of water when it is stirred
by paddles driven by a slowly falling weight. (Joule's success rested
on how good his thermometers were — he claimed an accuracy of
$1/200°$F.) The result was that 1 thermal unit centigrade is generated
by 1390 foot-pounds (ft lbs) of mechanical energy. Joule called this
relationship the *mechanical equivalent of heat*. The foot-pound used
to be the common unit for mechanical energy; nowadays, with the
metric system, it's the *Joule* (J).

Joule was an indefatigable experimenter. Kelvin used to tell the
story of how by chance he had met Joule and his new wife, Amelia, on
their honeymoon in the Alps. Apparently, Joule was carrying one of
his large thermometers. So that he could measure the temperature of
the water at the top of a waterfall and at the bottom. To measure the
temperature difference, if any. For lots of waterfalls. His enthusiasm
and dedication obviously impressed Kelvin. (Amelia's thoughts on
the subject are not recorded.)

Joule's value for the mechanical equivalent of heat allows us to
convert the Sun's heat output to the more familiar units of foot-
pounds of mechanical energy. It is 2784×1390 ft lb per sec per square
foot; that is,

$$\text{solar energy output} = 3{,}869{,}000 \text{ ft lb/sec/sq ft.}$$

This was a crucial number for Kelvin as he searched for the source
of the Sun's energy and ultimately its age.

Source of the Sun's radiant energy (1854)

So where does the Sun's radiant energy come from? If you were
asked that question, what would you say? In 1854, Kelvin gave
his first "definitive" answer, in an address to the Royal Society of
Edinburgh — ironically, the very same place where, only 70 years
before, the geologist James Hutton had argued in favor of an Earth
of indefinite age.

Kelvin offered the following three possibilities:

I. The Sun was a hot body, gradually radiating off its "primitive" heat and cooling down.
II. The Sun gets its radiated heat from chemical reactions continually taking place on its surface.
III. The Sun derives its radiant energy from meteoric matter falling in on its surface.

And which one is the correct answer? Kelvin confidently claimed that the third option was "the only one of all conceivable causes".

How did Kelvin come to this startling conclusion? As spelled out in his publication, it's a long and tortuous path. Interesting to follow — at least for a theoretical physicist like me. More than twenty different calculations. A theorist's paradise!

But, apart from yielding to temptation just once (in Appendix 3), I won't go there. After all, we are interested primarily in the Earth. So in what follows, I'll only give a brief description (with relevant numbers) of Kelvin's major steps for the Sun. His later, more important paper in 1862 is based on it.

Possibility I

In his 1854 paper, it took Kelvin only a few sentences to discuss — and dismiss — this first possibility.

Primitive heat

Think of the Sun as a *solid* body, radiating off its "primitive" heat. Means that heat energy has to be continually transferred from the inside outward to the Sun's surface — by *conduction*. A lot of it, exactly how much being determined by Pouillet. But the ability of ordinary materials to conduct heat is limited. Which restricts the outward flow of heat to the surface. So heat from greater depths just can't get to the surface fast enough to sustain the Sun's output. Which would mean that the heat being radiated by the Sun must come primarily from near its surface. According to Kelvin, the top inch or two! Which has the inevitable consequence that the Sun would soon become dark — "in two or three minutes, or days, or months, or years".

Definitely a non-starter.

In a footnote added later, Kelvin briefly considered the possibility that the Sun might be *molten*. Has the advantage that heat can be transferred to the surface much more effectively by *convection*. The key quantity then is what is called the *specific heat* — the amount of heat energy that 1 lb of material releases when its temperature is lowered by one degree. Kelvin showed that, for typical values of specific heats of materials found on Earth, the Sun's temperature would fall by 1.4 degrees centigrade in only 1 year. That's almost 3000 degrees centigrade in 2000 years! Astronomers would have noticed that. They hadn't. Another non-starter.

In fact, Kelvin felt he could rule out this liquid model of the Sun on another score. For cooling is usually accompanied by contraction. Kelvin estimated that, for the Sun to shrink by 1%, it would take only about 860 years. That's about 2.3% in 2000 years. Astronomers are not blind. They would have noticed that too.

A liquid Sun then? Definitely not. Kelvin ended up "utterly rejecting" it.

Possibility II

What about the Sun being a big ball of fire? Generating its own heat by combustion? Chemistry. It's possible. But how probable? Kelvin soldiered on.

Two lines of attack, depending on the material. Ordinary coal-like stuff. Or explosives. (It's hard not to get the impression that Kelvin himself had a lot of fun with these particular calculations — I've put them in Appendix 3.)

Net result — both ruled out. Chemistry not an option. Here's why.

The fiery furnace

First, what about materials like coal? The black gold that, in Kelvin's mid-19th century Victorian England, was fueling the steam engines of the Industrial Revolution. Remembered for the great wealth it generated (for a few). And its "dark, satanic mills" (for the many). The question was this — how much coal would need to be burning at the Sun's surface to generate enough radiant energy (known from Pouillet)?

The result? As Kelvin succinctly put it, the Sun's output was equivalent per square foot to the energy generated by "the fires of the whole [British] Baltic fleet, heaped up and kept in full combustion". The demand for oxygen would be huge, so huge in fact it wouldn't work. Conclusion — "the fire would be choked ... no such fire could be kept alight for more than a few minutes". No room for "fire worshippers" here!

The gunpowder plot

But not all materials need a supply of oxygen to combust. Explosives, like gunpowder, have their own internal supply of oxygen. Kelvin, leaving no stone unturned, considered this next.

To do this calculation, you need to know how much heat is generated when 1 lb of explosive detonates. Then compare that with Pouillet's number for the Sun's heat output. Kelvin concluded that it would need about 0.7 lb of explosive material being consumed at the Sun's surface per second per square feet to provide the Sun's output.

Any consequences? Yes, fuel consumption at the surface would cause the Sun to gradually diminish in size. How fast? About 0.009 ft in depth per sec. Doesn't sound like much. About 55 miles per year. Or 55,000 miles in a thousand years. Given that the Sun's radius is about 441,000 miles, Kelvin concluded that, about 8000 years ago, the Sun would have been twice as big as it is now. And it will be totally gone in another 8000 years.

No way. The Sun wasn't getting its energy from explosives. Or by chemistry of any kind.

Possibility III

Could a vast cloud of meteors be the answer? Is this the "truth" our Sherlock was after, no matter how improbable? Kelvin persevered.

There's certainly plenty of evidence for meteoric material out there in space. For instance — the "shooting stars" we see on a clear night. They're caused by dust particles burning up in the Earth's outer atmosphere. There's also a less well-known effect called "zodiacal light". It's a very faint glow of light that can sometimes be seen in the western horizon a few hours after sunset, and in the eastern sky a few hours before sunrise. Caused by the scatter of sunlight from particles out there in space between the Earth and the Sun.

So, there's definitely "stuff" out there. Enough to fuel the Sun? Need to calculate. Calculate — Kelvin was in his element.

Kelvin imagined (at least to begin with) a large cloud of meteors, stretching out to the edge of the solar system and beyond. Each meteor in this cloud is attracted toward the Sun. By Newton's Law of Gravitation, $F = GM_\odot m/r^2$, where $M_\odot$ is the mass of the Sun, m is the mass of the meteor, r is the distance between the Sun and the meteor, and G is Newton's gravitational constant. These meteors speed up as they head for the Sun, eventually crashing into its surface. At up to 390 miles per second. That's 1.4 million mph! The heat generated by these violent collisions was, according to Kelvin, the source of the energy that the Sun needed to keep it hot and radiate.

The big question of course was — could this mechanism deliver enough energy? Even so, was there any evidence to contradict this idea?

Gravitational potential energy from meteors

Up until now, Kelvin had used pretty much standard old-fashioned physics. To determine the effect of his meteors, however, he turned to a new technique. Had discovered it in an obscure journal when he was 21. An 1828 essay on *The Application of Mathematical Analysis*, by a George Green. A miller's son who had published it at his own expense. Sold a grand total of 51 copies. Mostly to sympathetic friends. Went on to Cambridge University afterwards as an *undergraduate*, at the age of 40. Died four years later. His work came to underpin much of modern mathematical physics. Not bad for someone with essentially no schooling! (Unfortunately, not everyone lives long enough to reap their richly deserved recognition.)

What Green had introduced was the concept of what is called a "potential function". From which one can derive an expression for what is called the "potential energy". In this case, gravitational potential energy — energy released by a meteor as it falls toward the Sun. All the time gaining kinetic energy.

For a meteor of mass m, the amount of gravitational potential energy released (and hence kinetic energy gained) by the time it reaches the surface of the Sun is given by

$$V(R_\odot) = GM_\odot m/R_\odot. \tag{4.1}$$

It is this energy which, at collision, is transformed into heat.

So, how much of this meteoric matter is needed to sustain the Sun's energy output?

Using the above formula for gravitational potential energy, Kelvin quickly deduced that 1 lb of meteoric matter coming from afar would deliver 6.5×10^{10} ft lb of energy at the Sun's surface. So in order to sustain the total solar heat output by this mechanism, the deposition rate (as follows from Pouillet) would have to be $3.869 \times 10^6/6.5 \times 10^{10}$ or 6.0×10^{-5} lb per sec per sq ft of the Sun's surface. Or as Kelvin put it — about 1 lb per sq ft every 5 hours.

Is this a reasonable number? Does it lead to any contradictions? Let's check — what about the overall accumulation of meteoric matter on the Sun's surface? It would mean that the Sun would be getting bigger. How fast? By about 30 ft per year. That's only 10 miles in 2000 years. Not observable.

So far, so good. Nothing to rule out this idea. Kelvin — never bashful — proudly proclaimed: "the source of the energy from which solar heat is derived is undoubtedly meteoric".

The case of the diminishing cloud

But wait. We're not quite finished. Kelvin soon found two problems. They caused him to retreat to a modified position.

First, those shooting stars. Caused by particles of matter streaking through the Earth's upper atmosphere. Burning up. You can count them — about 10 every hour. How many would you expect if they were caused by Kelvin's meteors rushing toward the Sun? The answer — a lot more than 10 per hour! (For once, Kelvin doesn't quote a number; I can understand why — it's several orders of magnitude too big!)

And second, more serious, the meteoric matter accumulating on the Sun would cause the Sun's mass to gradually increase. Which, because of the increasing gravitational pull of the Sun, would affect the orbital dynamics of all the planets. Including the Earth: its orbital radius around the Sun would decrease, its orbital speed would increase, and the time for each revolution would decrease. Now wind the clock back, say 2000 years. To Ptolemy. Events such as new and full moons, eclipses, etc. would all have occurred at slightly different times from what we would expect if the Earth had always been revolving around the Sun at its present rate. Different by about one and a half months. No shift like this is observed.

Looks now as if Kelvin is in trouble. Can his theory be saved? Yes, it can. The stability of the Earth's orbit must mean that the whole meteoric cloud must all be contained *within* the Earth's orbit. Gradually *spiraling* in toward the Sun. No problem now with an overcount of shooting stars. And the Earth's orbital dynamics remain stable, since the amount of matter within its orbit is constant. The revised theory is looking good.

To support this new model, Kelvin noted that the observed paths of shooting stars do tend to agree more with the picture of them spiraling in toward the Sun than heading straight toward the Sun. One slight drawback however is that these spiraling meteors will deliver less energy (half the amount) than before. So you need twice as many to crash into the Sun to keep the Sun going. Leading to an increasing depth on the surface of 60 ft per year (not 30 ft per year as before). That's about 20 miles in 2000 years. Still too small to be observed. So not a problem.

This was Kelvin's ultimate theory in 1854 for the maintenance of the Sun's energy output. A meteoric cloud lying entirely within the Earth's orbit. Gradually spiraling downward toward the Sun. Explained everything.

Though it didn't convince the geologists.

The demise of the meteoric cloud

By 1862, things had changed. A lot. Kelvin's meteoric cloud was dead. How did that happen? And was there anything he could do to rescue the situation?

It was astronomy that killed Kelvin's swirling cloud of meteors. Kelvin himself had recognized that his meteor theory wasn't infallible. For his cloud of meteors, lying within the Earth's orbit, would disturb the orbits of the two innermost planets, Mercury and Venus. Sure enough, in 1859, an unexplained discrepancy in Mercury's orbit was discovered, by the French astronomer Urbain J.J. LeVerrier. Has to do with the axes of Mercury's elliptical orbit. They slowly revolve in space, a characteristic known as a "precession". Caused by the influence of the big planets Jupiter and Saturn. Trouble was — Mercury's orbit was precessing at the wrong rate (by an amount of 38 seconds of arc per century) in excess of the calculated rate.

Good news for Kelvin? Could his reduced meteoric cloud be the source of this unaccounted-for piece of the precession? Do the calculations. Answer — no. There was just too much meteoric material in his cloud (determined, as you may recall, by how much material is needed to keep the Sun ablaze). So Kelvin's cloud got eliminated by the smallness of LeVerrier's number.

One last gasp. What if the entire meteoric cloud lay within the orbit of Mercury? Sorry. That won't do either. Kelvin himself quickly realized that such a dense cloud of meteoric matter would perturb the orbits of those comets that pass close to the Sun. No such effect had ever been seen.

Conclusion — the Sun wasn't sustained by a host of meteors after all. Kelvin had previously eliminated his Possibility I and Possibility II. Now his only other candidate, Possibility III, had likewise been eliminated!

(LeVerrier's detection of an unexplained component in the precessional rate of Mercury's orbit was to play an extremely significant role 56 years later. It provided a confirmation for new revolutionary physics — the unity of space–time and Einstein's General Theory of Relativity in 1915.)

Kelvin's last stand (1862)

A bit embarrassing. Ruling out all of your options. What was our master detective to do? Backtrack. There had to be a loophole somewhere.

Kelvin found it in Possibility I. Which had to be revised. He decided he had been too cavalier in dismissing the possibility of the Sun being *liquid*. Had previously rejected it because it led to a much too rapid cooling of the Sun. Around 3000 degrees centigrade in only 2000 years. And the shrinkage — more than 2% in those same 2000 years.

But in these earlier calculations, Kelvin had assumed that the values of various properties (specific heat, etc.) of the material in the Sun were similar to those found on Earth. What if their values in solar material were very different? Say greater by a factor of 10, or even 10,000! That would reduce the rate at which the Sun changes by a corresponding factor. Its temperature would decrease much, much more slowly. And the accompanying contraction would likewise be very much smaller. Previous perceptible changes now removed. Now imperceptible.

So yes, the Sun could be fixed. Again. Now a big ball of incandescent liquid. With convection. But at a price — the materials out of which the Sun is made had to possess unbelievably extraordinary properties.

And there was a source of energy that Kelvin had not previously considered. Maybe that should be included. It had been suggested to him by his good friend Hermann von Helmholtz. Gravitational contraction and compactification even after the Sun had been formed. A contraction of just 0.1%, for example, in the solar diameter could (assuming uniform solar density) provide enough energy to last 20,000 years. And a lot more when the density was much bigger, such as expected toward the center. Surely that too would contribute to providing radiant heat for the Sun. (Kelvin didn't follow up on this idea.)

The age of the Sun

So much for the origin of the Sun's radiant energy. What about the *formation* of the Sun itself?

There was of course always the possibility that it had been "created ... by an over-ruling decree". Not an idea that someone like Kelvin — a staunch Scottish Presbyterian — dismissed outright. Still, he thought that, if he could find an explanation "not contradictory to known physical laws", then that was the more likely one.

His own conclusion? Some kind of accretion. At least partly due to his beloved meteors (an idea he was reluctant to let go of). More likely, the coalition of many small bodies (another idea of Helmholtz). Nowadays called planetesimals.

So during the Sun's formation, gravitational potential energy from infalling planetesimals would be released, ultimately generating a huge supply of heat. How much heat? Roughly an amount given by the formula

$$V(R_\odot) = kGM_\odot^2/R_\odot, \qquad (4.2)$$

where as, before $M_\odot$ and $R_\odot$ are the resulting mass and radius of the Sun. The coefficient k is a constant that depends on how density varies with depth within the Sun. For example, if the density is uniform (as Kelvin assumed), then $k = 3/5$. (This formula can be derived using calculus, or more directly by applying what is called "dimensional analysis" — a particularly simple technique introduced in 1822 by the French mathematician Jean-Baptiste Joseph Fourier, of whom we shall soon hear much more.)

How much energy is this? Plug in the numbers. The result is

gravitational potential energy released = 15.3×10^{40} ft lb.   (4.3)

That's a lot of energy! Compare it with the annual energy output of the Sun (8.32×10^{32} ft lb per year, from Pouillet). Which tells us that if, in the past, the Sun had been radiating energy at about the same rate as it is now, there's enough energy there to maintain such an output for about 20 million years. Or more if the Sun's density increased toward its center. Kelvin's upper limit was 500 million years.

So that's it. Kelvin's estimate for the age of the Sun. Some of the assumptions a bit dubious maybe. But Kelvin was totally convinced that the Sun had been shinning for a limited time of between 20 million and 500 million years. Ballpark figure — 100 million years. A much shorter timescale than the one demanded by his geological and biological colleagues. They wanted much, much more time. A stable Sun. Providing the Earth with a stable environment.

Disagreement. And matters weren't improved by Kelvin's next publication a month later — an estimate of the age of the Earth. By a totally different method. Same answer — 100 million years. To Kelvin, confirmation.

The age of the Earth

Having dispensed with the Sun, Kelvin turned his attention to the Earth. There, he was on firmer ground. No question about the materials it was made of. And by the 1860s, many of their thermal properties — such as their melting temperatures, their thermal conductivities, and specific heats — had all been measured. Now was the time to put the numbers together with the mathematics. Kelvin was the one to do it, to calculate — by irrefutable means — how old planet Earth was.

Kelvin had had a long-standing interest in this question. And he was up-to-date with his mathematics. As a youth, he had gone to the right school — Glasgow University. It was an unusually enlightened place. His family had moved to Glasgow from Belfast in 1833 when his father was appointed Professor of Mathematics. The precocious Kelvin (then just William Thomson) was encouraged to learn all he could about the latest developments in mathematics. Especially the works of the great continental mathematicians — Euler, Lagrange, Laplace, Legendre, and Fourier. All inspired by Leibniz's flexible analytical approach to calculus. A big contrast with his experiences at Cambridge where he transferred in 1841. There the emphasis was on problem-solving. Lots of it. To relieve the drudgery, Kelvin launched himself (with typical energy) into sculling. Won the Colquhoun Trophy there in 1843. Also helped form the University Musical Society, playing the cornet and French horn. It was Kelvin's knowledge and understanding of a new mathematical approach that was to help him find an estimate for the age of the Earth.

The Earth was of course a specific example of a thermodynamical system. Kelvin imagined that, in times long past, the Earth had once been molten. Very hot. But cooling, with heat energy being radiating off into cold space. Like the Sun. At some point, the Earth would solidify. Right to the core. Cooling would continue, the Earth's heat being gradually dissipated into outer space. Irreversibly.

The question then was — how long would it take for the Earth to cool down from the time it first became a hot solid to its present much cooler state?

Fourier's law for heat conductivity

Kelvin was lucky. The mathematics had already all been done. By Jean-Baptiste-Joseph Fourier in France. In his great treatise *Théorie Analytique de la Chaleur* of 1822. And Kelvin was very familiar with it. He had read it as a teenager, surreptitiously devouring its pages instead of taking French lessons with his sisters. Had even published two short papers defending Fourier's work. Kelvin was already famous by the time he arrived as a student at Cambridge!

Fourier's theory of heat conduction was based on four principles:

(1) Heat is directional — it flows from higher temperatures to lower temperatures. This doesn't say what heat really is. It was originally thought to be some kind of fluid — "caloric"; our use of the words "heat flow" is a lingering reminder of that concept. It was only later in the 19th century that physicists began to realize that heat conduction was related to molecular vibrations.

(2) Heat is conserved. This is a special case of the more general law of conservation of energy which includes all possible kinds of

energy — a law that Kelvin (with his friend Joule) later helped to formulate.

(3) How much heat flows in any given situation depends on the material's ability to transfer the heat — a property known as its *thermal conductivity*. A material that is able to conduct heat more easily than another has a higher thermal conductivity.

(4) Finally, the amount of heat that flows through a given region in a material is governed by how much the temperature is changing in that region. That is, the temperature gradient.

Fourier derived a general equation which describes this heat flow. Here, we need to only consider heat flowing in just one direction — a big simplification, but that's all Kelvin needed to find his estimate for the age of the Earth.

Suppose we focus on a point P as shown in Figure 2. It is at depth x below the surface of the Earth.

The temperature there we'll denote by $T(x, t)$. As indicated, it depends not only on x (the depth), but also on the time t (since the

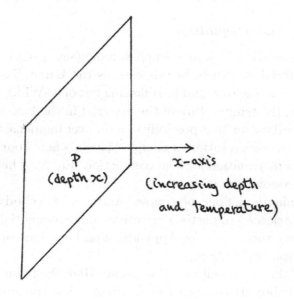

Figure 2: Plane across which heat is flowing.

temperature at depth x can — and does — change with time as the Earth cools).

We've also drawn through P a plane perpendicular to the x-axis. Since the temperature of the Earth increases with depth, the temperature on the right-hand side of this plane is higher than the temperature on the left-hand side. So as the Earth cools, heat will flow through this plane from right to left.

Now let $H(x,t)$ denote the amount of heat flowing (from right to left) through unit area of this plane, per second. And $G(x,t)$ the temperature gradient at depth x and time t.

Then, according to Fourier, H and G are related by

$$H(x,t) = \sigma G(x,t), \qquad (4.4)$$

where σ represents the thermal conductivity of the material. This equation expresses the property that the heat flow (per unit area per second) is proportional to both the thermal conductivity and the temperature gradient. It's called Fourier's Law for Heat Conduction.

Fourier's master equation

Next, let's take a look at a small volume (say a box) of the conducting material — which, in this case, is the Earth. There will be heat flowing into the box, and heat flowing out of it. Which needn't be equal. If not, the temperature of the material in the box will change. We've got to allow for that possibility. Whatever happens, it's all constrained by the conservation of energy. Which is how Fourier derived his fundamental equation for heat conduction (Eq. (4.5) below). After that, find the solution.

Easier said than done of course. And not everybody wants to follow the intricacies of partial derivatives and exponential functions. So, I'll reserve the details for Appendix 4 and continue on here with a more general description.

Figure 3 shows a small box of material. Here the point P is at the *center* of the box. It is located at depth x below the surface of the Earth. $T(x,t)$ is the temperature there at time t.

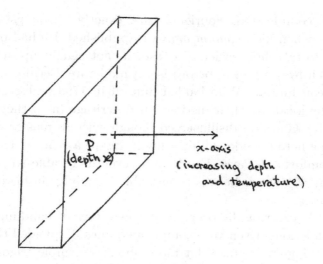

Figure 3: Small box of material.

What about the thermodynamics?

Since it's cooler at the surface of the Earth than at its center, heat is flowing from right to left in this diagram. So heat is flowing *into* the box through its right-hand face, and *out* through its left-hand face. Any difference between the two will cause the temperature of the material in the box to change. At all times, heat energy is conserved.

All of which leads (see Appendix 4) to Fourier's equation for heat conduction. This is what it looks like:

$$\frac{\partial}{\partial x}G(x,t) = \frac{1}{\kappa}\frac{\partial}{\partial t}T(x,t), \qquad (4.5)$$

where κ is a quantity that Kelvin called the *diffusivity* of the material. As before, $G(x,t)$ is the temperature gradient at depth x and time t, the temperature gradient itself being $G(x,t) = \frac{\partial}{\partial x}T(x,t)$. The expression $\frac{\partial}{\partial x}G(x,t)$ indicates the rate at which this temperature gradient is changing with depth. Finally, the symbol $\frac{\partial}{\partial t}T(x,t)$ denotes the rate at which the temperature at depth x changes with time t.

Pretty complicated. Fourier wasn't exactly a youngster — he was 54 — when his *magnum opus* was published. He had originally intended to take holy orders. Instead, he got caught up in politics. The French Revolution to be precise. A moderate. Became a favorite of Napoleon himself. Who hauled him off in 1798 to Egypt to lecture to the locals on their distinguished heritage (as if they weren't aware of it). Got a bit disillusioned though with his master when the conquering hero abandoned his army in Egypt a year later to return to the comforts of Paris. Fourier was eventually able to return to France in 1801 where he was at last able to indulge in his first love, mathematics.

Fourier's greatest discovery was a new way of handling mathematical functions. Even very complicated ones. Expanded them as a series of things we know a lot more about — simple trigonometric functions (sines and cosines). Makes life a lot easier. And you can solve all sorts of problems that had hitherto been intractable.

All of which came to light in 1811 when Fourier submitted a memoir to the French Institute in Paris for that year's Grand Competition. The subject — the propagation of heat through solid bodies. Set by an illustrious committee that included Lagrange and Laplace. They were impressed. But, with some reservations — not rigorous enough! (A favorite mantra of mathematicians.) Fortunately however, good enough for Fourier to be awarded the Prize. Still, it took another 11 years before his work was fully accepted and published.

In the years that followed, Fourier's approach has opened up a whole new way of tackling problems, not just in physics, but essentially in every area of science and technology. You might know it by the words "Fourier series" and "Fourier transforms". Indispensable in today's modern world.

The cooling Earth

Kelvin saw the cooling Earth as just the very place where he could apply Fourier's methods.

Start off (at time $t = 0$) with the Earth having just turned solid. Its temperature is T_0. As time progresses, the Earth's internal heat is conducted to the surface where it is radiated away. Question — how long would it take for the Earth to arrive at its present condition? All Kelvin had to do was apply Fourier.

But first, let's make a few assumptions to simplify life. Like neglecting the curvature of the Earth. So, heat flows outward essentially one-dimensionally. We'll also take the diffusivity κ to have the same value throughout the body of the Earth. Makes the mathematics a lot easier.

It's just a model, of course. That's what physicists do when the going gets tough. Try something simpler. Might give some guidance. Approach reality in steps. Which is what Kelvin was doing here. Even so, it's still too hard to get a nice, neat formula for what we want — the temperature $T(x,t)$ for all depths x and time t. The best you can do is to express $T(x,t)$ in a complicated integral form which can only be evaluated numerically.

But lo and behold, it is possible to extract a simple formula for the *temperature gradient* $G(x,t)$ (see Appendix 4). And from this, an even simpler expression for the temperature gradient at the Earth's *surface* — one of the few characteristics of planet Earth that can actually be *measured*.

The formula for the temperature gradient at the surface is

$$\text{surface temperature gradient} = T_0/\sqrt{\pi\kappa t}. \qquad (4.6)$$

Here, t is the time it has taken for the Earth to achieve this value of the surface gradient since its formation as a hot solid.

Just one more step. Re-arrange the above equation to solve for the time t:

$$t = \frac{1}{\pi\kappa}\left(\frac{T_0}{\text{surface temp. gradient}}\right)^2. \qquad (4.7)$$

An amazingly simple result. It means that, if you know the temperature gradient at the Earth's surface *now*, you can determine from this formula (assuming T_0 and κ are known), the amount of time t that has passed since the Earth first solidified. That is, the geological age of the Earth.

Kelvin's 100 million years

Ever since his inaugural address at Glasgow in 1846, Kelvin had been urging his fellow scientists to do a geothermal survey to find how the Earth's temperature changed with depth. It was well known that temperature increases with depth — ask any coal miner. But *exactly* how much did it vary with depth?

Kelvin was always a man of action. A do-er. A leader — working closely, for example, with his friend James Forbes in Edinburgh. Measuring as accurately as possible how temperature changed with depth at three nearby locations. For 20 years or more!

There are of course two well-known regular temperature fluctuations near the Earth's surface. Both short-term. The day/night effect, due to the Earth's daily rotation. And the summer/winter effect, due to the Earth revolving in its orbit around the Sun. But these penetrate only about 3–4 ft and 60–70 ft, respectively. Below this depth, the temperature doesn't show these variations. It just increases steadily the deeper you go. This is the temperature gradient we want. And though not the same everywhere, Kelvin made a rough estimate of about 1°F increase in temperature for every 50 ft in depth.

For the initial temperature of the just-solidified Earth, Kelvin took $T_0 = 7000°F$ (3871°C). And for the diffusivity (taking average values of thermal conductivity, specific heat, and density of ordinary materials), he took $\kappa = 400\,\text{ft}^2/\text{yr}$.

Now put it all together. What did Kelvin get for the age of the Earth? We have

$$t = \left(\frac{1}{400\pi}\frac{\text{yr}}{\text{ft}^2}\right)\left(\frac{7000°F}{1°F/50\,\text{ft}}\right)^2 = 98 \text{ million years.} \qquad (4.8)$$

Not 6000 years. Not 300 million years. And certainly not infinity.

Kelvin of course realized the uncertainty of some of the numbers he had used. In particular, it was a bit far-fetched for the diffusivity to be constant throughout the Earth. Surely it would vary with internal pressure and temperature. Allowing for this, Kelvin confidently put the age of the Earth in the range between 20 million and 400 million years. Physics had decided it. "Elementary, my dear Watson!"

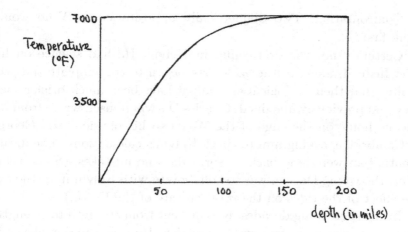

Figure 4: Temperature profile in Kelvin's model of the Earth.

One more point about Kelvin's model of the Earth — its temperature profile. How the temperature varies with depth. It's shown in Figure 4.

What's revealing here is that, in Kelvin's model, the internal temperature of the Earth rises very sharply with depth. In only a little over 100 miles (the radius of the Earth is more than 4000 miles), the temperature has skyrocketed to almost 7000°F, Kelvin's original temperature for the whole Earth. Which means that, in all the time that Kelvin's Earth has been cooling, most of the energy that was radiated away by the Earth, came from just the top sliver — a hundred or so miles! Very little from its inner depths. For 100 million years.

Denouement

So there it was, in his 1862 paper, Kelvin's historic estimate for the age of the Earth since it first solidified. Based on physics. The incontrovertible laws of thermodynamics. It limited how old the Earth could be. About the same age as the Sun. Surely that in itself was confirmation!

So all those "extreme quietists" (as Kelvin called the geologists and biologists) were wrong. The vast amounts of relatively stable, peaceful time they demanded were impossible.

Confrontation. Two diametrically opposite views. Who would blink first?

Certainly not the ever-confident Kelvin. He had science on his side. Instead, many of the geologists began to contemplate the possibility that their geological eras might have been much briefer than they had previously imagined. Charles Darwin himself wrote from his bucolic home on the edge of the Weald to his physicist son George at Cambridge, asking him to check Kelvin's calculations. The disappointing answer came back — sorry dad, no mistakes. (Some years later, the young Darwin was to collaborate with Kelvin in a study of the effect of the tides on the rotation rate of the Earth.)

In the succeeding decades, Kelvin went from strength to strength, becoming even more famous — and rich. He was a major player in the ultimate success of the transatlantic telegraph cable, finally completed in 1866. Which included the use of Kelvin's receiver, the only one sensitive enough to pick up the signals sent out from the other end of the line. (His contributions were immediately recognized with a knighthood — Sir William.) No stopping his inventiveness. Eventually, he ended up with 70 patents! Including a marine compass which compensated for errors arising from a metal ship's own magnetism. And a tidal gauge, which could predict tides (a modern version of which was used on D-day in 1944 to time the Allied landings on the Normandy coast).

Those who expressed legitimate criticisms of Kelvin's estimate for the age of the Earth felt they had run into a brick wall. Such as John Perry in 1895. Had once been a student of Kelvin's. Now he was pointing out a serious flaw in his mentor's arguments. Kelvin had assumed that the Earth was solid, throughout. What if the inner Earth was liquid? That would change things a whole lot. For convection would then be possible. And convection is a much more effective way of conveying heat than conduction. Makes internal heat much more available at the surface. So, it would take longer — possibly much longer — for the surface temperature gradient to decrease from its original value to its present value of about 1°F per 50 ft. Perry's calculations in fact showed that the Earth could easily be a billion years old. Or more. Which made little impact on the debate.

But Kelvin's model did eventually fall. Several years after the discovery in 1896 of the new phenomenon of radioactivity by Henri Becquerel in France. For in 1903, Pierre Curie and his assistant Albert Laborde made the important discovery that the radioactive element radium releases energy when it decays. Which eventually becomes heat. Likewise for other radioactive elements such as uranium and thorium. Meant that one of Kelvin's basic assumptions — that there was no heat source within the Earth itself — was wrong. His model undermined. His estimate for the age of the Earth untenable. With a continuous internal source of heat, the Earth could be much, much older than Kelvin ever imagined. Ecstatic relief from geologists and biological evolutionists!

Of course, it wasn't known how much radioactive material there was within the Earth. Still isn't. Scientists nowadays make educated guesses as a part of their now very sophisticated computer-generated models of the Earth. And it will be many years more before the amount (and distribution) of these radioactive materials within the Earth is known. Most likely by measuring the energy and flux of what are called anti-neutrinos — by-products of certain radioactive processes which can pass with ease through the Earth's material to the surface where they can be studied.

Perry however had a large piece of the truth. The internal structure of the Earth is much more complicated than Kelvin ever imagined. There's a solid "mantle", going down to a depth of about 1800 miles (where the temperature is indeed about 7000°F). And at the center of the Earth, a solid "inner core" whose outer radius is about 760 miles (the temperature there being about 9000°F). Between the inner core and the mantle, there's an "outer core" which is liquid. Where convection can and does take place.

All of which was revealed in the twentieth century. By clever seismologists. Aided by computers with vast number-crunching powers.

And there's no way that Kelvin could have imagined what the Sun is like. Not a solid, not liquid, but a big ball of *gas*! Right to its very core. Temperatures ranging from about 5800 degrees on Kelvin's absolute temperature scale at the surface to 15.8 million degrees at the center. Mostly hydrogen. With some helium, and traces of other elements. Powered by a process totally unknown in the 19th century. Nuclear fusion. Hydrogen being turned into helium. Mass into energy. Formed about 4.6 billion years ago. Out of a vast molecular cloud, self-gravitating. And which will come to an end in another 5 billion years or so when the hydrogen fuel at the core begins to get depleted. Followed by a series of convulsions. Which throw off its outer layers. Leaving as a remnant the Sun's hot inner core which, like the old soldier, gradually fades away. To unspectacular oblivion.

Kelvin though was right on one aspect — it's all based on physics. Trouble was — neither he nor anyone else in his day knew the right physics.

For the rest of his life, Kelvin remained popular. Feted wherever he went. Raised to the peerage in 1892 — Lord Kelvin of Largs. After the River Kelvin which flows past his beloved university, and Largs where his substantial home was (on the Ayrshire coast south of Glasgow). He retired from teaching in 1899 after more than 50 years. I remember sitting as a freshman in the same cavernous lecture room — in the same Department of Natural Philosophy — where Kelvin had taught many years before. Tier after steep tier of dark wooden benches that seemed to merge into the distant darkness of the ceiling. Interestingly though, in a corner up front, hung a large weighted pendulum (the bob being a metal pipe packed with lead, with a wooden plug at one end). Apparently Kelvin used to fire an elephant gun at it, to measure the speed of the bullet from the resulting period of the swing. Don't have such dramatic lecture demonstrations anymore! Not allowed. Health and safety — against the rules.

Something else that's different nowadays — Kelvin used to start each of his lecture sessions with a prayer (all students standing). Apparently always the same one — the Third Collect (for Grace) in the Book of Common Prayer (... "grant that we fall into no sin" ...). I wonder what he had in mind.

But some things haven't changed. Such as laboratories for students to gain hands-on experience. Kelvin was the first to initiate these in the whole of Great Britain. And Kelvin would recognize the contents of many of the modern physics textbooks — they're

all based on the definitive textbook that he wrote with his friend
Peter Tate (the book was known fondly by its many users as T and
T′). It was the first textbook to emphasize the importance of conservation laws. All physics textbooks, without exception, do the same
nowadays.

Apparently Kelvin had his own way of keeping his students on
their toes. Using a box with three compartments. At the beginning
of each term, the names of all his students were placed in the center
compartment — known as Purgatory. Every now and then, Kelvin
would pause, pick out a name and ask the student a question related
to the topic under discussion. If the student answered correctly, the
name was returned to one of the other boxes — known blissfully as
Heaven. If however the answer was wrong, the student's name would
go into the remaining box — known as Hell. To be tormented at a
later date with more questions. (Appeals to my teaching instincts as
not such a bad idea!)

Kelvin died peacefully in 1907, aged 83. At his home in Largs.
Memorial services followed. First at the local St. Columba's Kirk
where he had been an elder, then at Glasgow University. Finally laid
to rest, after much ceremony, in Westminster Abbey in London. Near
Sir Isaac himself.

In his day, Kelvin was revered, considered by many to be the
greatest scientist of the 19th century. But alas, times change. And
with it comes re-evaluation. Kelvin no longer holds this honor.
Instead, it has passed to his good friend and fellow Scot James Clerk
Maxwell. Without whom there would be no cell phones, no iPods,
no laptops, no television, no GPS, no. . . . You get the idea. Maxwell,
not Kelvin.

Appendix 1: Numerical values, etc.

Below is a list of various numerical values that Kelvin used in his calculations. Modern values are slightly different, but the conclusions are unchanged.

Weight of the Sun	4.23×10^{30} lb
Radius of the Sun	441×10^3 miles (232.8×10^7 ft)
Weight of the Earth	119.2×10^{23} lb
Radius of the Earth	3956 miles (2.09×10^7 ft)
Radius of Earth's orbit around the Sun	95×10^6 miles
Radius of Mercury's orbit around the Sun	36.8×10^6 miles
Acceleration due to gravity at the Earth's surface g	32.2 ft/sec^2
Acceleration due to gravity at the Sun's surface	28 g

Weight = (mass) g

Surface area of a sphere of radius R is $4\pi R^2$

1 cubic ft of water weighs 62.4 lb

Kelvin took 1 cubic ft of solar material to weigh ($62.4 \times 1\frac{1}{4}$) lb or 78.0 lb

1 degree centigrade = 1°C

1 thermal unit centigrade = 1390 ft lb

1 year = 3.156×10^7 sec

1 degree = 3600 seconds of arc = $\pi/180$ radians

Appendix 2: The Sun's heat output, from Pouillet

Pouillet measured the amount of heat energy received from the Sun per second per square foot at the Earth's surface. Kelvin quotes a value of 0.06 thermal units centigrade per second per square foot.

The beauty of this number is that it can be turned into information about the *Sun* — the rate at which it is radiating energy. For all the energy being radiated from the Sun (a spherical surface

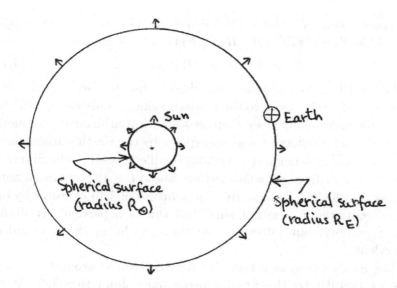

Figure 5: All the heat energy radiated outwards from the Sun's surface passes through a sphere whose radius is the radius of the Earth's orbit around the Sun.

of radius $R_\odot$) must pass through the much larger spherical surface whose radius is R_E, the radius of the Earth's orbit around the Sun (see Figure 5 which is the same as Figure 1). Thus,

(energy radiating per sec per sq ft from Sun's surface)$(4\pi R_\odot^2)$

= (energy incident per sec per sq ft on a sphere of rad R_E)$(4\pi R_E^2)$,

(4.9)

where $4\pi R^2$ is the surface area of a sphere of radius R.
Using Pouillet's measurement, we find that

(energy radiated per sec per sq ft of the Sun's surface)

$= (0.06)(R_E/R_\odot)^2$ thermal units centigrade per sec per sq ft

= 2784 thermal units centigrade per sec per sq ft

$= 3.869 \times 10^6$ ft lb per sec per sq ft. (4.10)

One more number to go — the rate at which energy is radiating not just from 1 sq ft but from the entire surface of the Sun. Otherwise known as the Sun's *luminosity* $L_\odot$. All we need to do is to multiply the above value of 3.869×10^6 ft lb per sec per sq ft by the surface area of the Sun ($4\pi R_\odot^2$ with $R_\odot$ in feet). The result is

$$L_\odot = 2.64 \times 10^{26} \text{ ft lb per sec}, \qquad (4.11)$$

or 8.32×10^{33} ft lb per year. Amazingly, this number for the solar luminosity is pretty close to the modern value — only about 10% too low. A bit lucky really since Pouillet somewhat arbitrarily assumed a factor of 1/3 for reflection and absorption by the Earth's atmosphere.

Kelvin had another way of stating Pouillet's result: "the Sun radiates every year from its whole surface about 6×10^{30} times as much heat as is sufficient to raise the temperature of 1 lb of water by one degree centigrade". I'm not sure that this is a particularly illuminating number, but (after you go through the math) it is indeed equivalent.

The Sun's luminosity had also been measured around the same time as Pouillet by the English astronomer John Herschel. At his observatory in Cape Town, South Africa. Normally busy cataloging the stars, nebulae, etc. in the southern heavens. Herschel measured the rate at which ice melts in sunlight — one inch in 2 hr 12 min 42 sec. Which translates to a number which is about 20% higher than Pouillet's. Kelvin seems to have stuck throughout his own calculations with Pouillet's result.

Appendix 3: Energy from chemistry?

Can the Sun generate its radiated heat by chemistry? Kelvin considered two different kinds of processes that could in principle do this. Involving materials like *coal* which require an external supply of oxygen for combustion. Or like *gunpowder* which have enough oxygen contained within themselves and which explode on detonation.

• What about *coal*? The relevant number here is that, on combustion, 1 lb of coal generates about 6600 thermal units centigrade. Which is equivalent (following Joule) to about 9.2×10^6 ft lb of mechanical energy.

Now compare this with Pouillet's measurement of the Sun's heat output. It means that it would need about 0.42 lb of coal to be consumed per second per square feet, burning at the Sun's surface, to provide all of its radiated heat.

To illustrate what this implies, Kelvin put his answer in terms of the ubiquitous steam-powered engines of his day. For he knew that, at best, coal burned in the furnaces of steam-driven locomotives and ships at a rate of 1 lb in 30 to 90 sec per sq ft of their gratebars. That's about 0.01–0.03 lb per sec per sq ft of a gratebar. So, you'll need 14–42 layers of gratebars (hence, his reference to the ships of the Baltic fleet being stacked on top of one another!) to generate the same amount of energy as the Sun puts out. An unlikely scenario. There's no way you could get enough oxygen into such a system to keep it ablaze for any length of time.

- Next, what about *explosives*? Like gunpowder (whose major oxygen-bearing component is saltpeter, which is potassium nitrate). Could the Sun get its energy from a material like this? Again, Kelvin knew the key number — 1 lb of explosive can generate about 4000 thermal units centigrade. Compare this with Pouillet. Result? It would require about 0.7 lb of explosive to be consumed at the Sun's surface per second per square feet.

Any consequences? Yes. With all this material being consumed at the Sun's surface, the Sun would be shrinking. At what rate? And would it be noticeable?

To calculate this, we need to know the *density* of solar material. An estimate for this can be obtained by dividing the Sun's weight by its volume $\frac{4}{3}\pi R_\odot^3$. The answer, as Kelvin states it, is about $1\frac{1}{4}$ times the density of water — that is, about 78 lb per cubic ft. It follows therefore that, for each square feet on the Sun's surface, you need only a thickness of 0.009 ft to yield a weight of 0.7 lb.

Looks pretty small. But multiply up. It corresponds to about 55 miles per year, or about 55,000 miles in 1000 years, or 440,000 miles in 8000 years. Which is about the whole radius of the present Sun. Kelvin's comment — the Sun would have been twice as big 8000 years ago, and will all be consumed in another 8000 years. Not likely!

Kelvin's confident summary — "the Sun does not get its heat by chemical action".

Appendix 4: Fourier's mathematics

Fourier's equation

Kelvin used Fourier's theory for heat conduction in his study of the Earth's cooling process. First, let's see how Fourier derived his famous equation. Then take a look at its solution.

We won't work out the most general form of this equation. Just the case where heat flows in only one direction. That's all Kelvin needed, the Earth's heat flowing directly toward its surface.

To do this, we're going to look at the balance of heat energy in the vicinity of a point P at depth x below the Earth's surface (which corresponds to $x = 0$). In particular, the heat balance for a small rectangular box (as shown in Figure 6) which has the point P at its *center*. The full width of the box is Δx.

The temperature at depth x at time t we'll denote by $T(x, t)$, and the temperature gradient there at time t is

$$G(x, t) = \frac{\partial T(x, t)}{\partial x}. \tag{4.12}$$

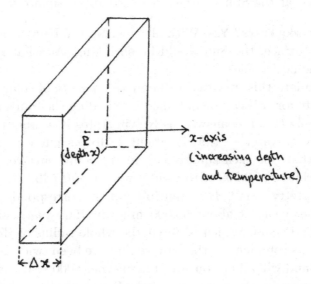

Figure 6: Small box of the Earth's material with heat flowing through it.

Since the Earth's temperature increases with depth x, the temperature gradient $G(x, t)$ at x is positive.

What we are going to do next is to look at the right-hand face of this little box. Its area is A. Since heat is flowing outward toward the Earth's surface (at $x = 0$), heat is flowing from right to left into the box. Exactly how much heat — which is what we are interested in — is determined by the temperature gradient at this face. Which is slightly different from what it is at the center x of the box (a distance $\Delta x/2$ to the left of this face). This is the tricky part: the temperature gradient at the right-hand face is given by

$$G + \frac{\partial G}{\partial x}\left(\frac{\Delta x}{2}\right), \tag{4.13}$$

which follows from taking the first two terms of what mathematicians call a Taylor series, after a Mr. Taylor who lived in England in the early 18th century.

Fourier's Law now tells us the amount of heat flowing in a brief time Δt through the right-hand face *into* the box is given by

$$H_{\mathrm{rh}}(A)(\Delta t) = \sigma\left[G + \frac{\partial G}{\partial x}\left(\frac{\Delta x}{2}\right)\right]A(\Delta t), \tag{4.14}$$

where H represents the heat flow per second per unit area.

Likewise, we can go through similar steps for the left-hand face which is at distance $(-\Delta x/2)$ from P. Heat is also flowing from right to left through this face, but now corresponds to heat flowing *out* of the box. From Fourier's Law, the amount of heat flowing through the left-hand face out of the box in time Δt is given by

$$H_{\mathrm{lh}}(A)(\Delta t) = \sigma\left[G + \frac{\partial G}{\partial x}\left(-\frac{\Delta x}{2}\right)\right]A(\Delta t). \tag{4.15}$$

Subtracting these two expressions tells us that the *net* amount of heat flowing *into* the box in time Δt is

$$\sigma\left(\frac{\partial G}{\partial x}\right)(\Delta x)A(\Delta t). \tag{4.16}$$

What happens to this heat? It's going to raise the temperature of the material in the box. By an amount ΔT say. The corresponding amount of heat can be expressed as

> (mass of material in box) (specific heat of material)
> (change of temperature).

That is,

$$\varrho A(\Delta x)s(\Delta T), \qquad (4.17)$$

where ϱ is the density of the material in the box, $A(\Delta x)$ is the volume of the box, and s is the specific heat of the material.

Equating (4.16) and (4.17), we find that

$$\frac{\partial G}{\partial x} = \frac{1}{\kappa}\frac{\partial T}{\partial t}, \qquad (4.18)$$

where $\kappa = \sigma/\varrho s$. Kelvin called κ the diffusivity of the material.

Following Fourier, since $G = \partial T/\partial x$, we can take this equation one step further, to get an equation entirely in T:

$$\frac{\partial^2 T}{\partial x^2} = \frac{1}{\kappa}\frac{\partial T}{\partial t}. \qquad (4.19)$$

This is Fourier's celebrated equation for heat conduction.

Solution

Mathematicians call this a second-order partial differential equation. Big words. Not always easy to solve. Certainly not in this case. In fact, there's *no* exact solution for $T(x, t)$ The best you can do is to write it in integral form:

$$T(x, t) = T_0 \cdot \frac{2}{\sqrt{\pi}} \int_0^{x/2\sqrt{\kappa t}} \exp(-u^2)du, \qquad (4.20)$$

where exp denotes the exponential function. Pretty horrendous!

By taking the derivative of this, we get for the temperature gradient at time t

$$G(x, t) = \frac{\partial T(x, t)}{\partial x} = \frac{T_0}{\sqrt{\pi \kappa t}} \exp(-x^2/4\kappa t). \qquad (4.21)$$

Looks as if we are getting into a morass of mathematics. We are! But luckily, we can bring it to a quick close. We're interested in the *surface* temperature gradient, that is, the temperature gradient when $x = 0$. But when $x = 0$, the exponential function in (4.21) reduces to 1. Hence, we can deduce that

$$\text{surface temperature gradient} = T_0/\sqrt{\pi \kappa t}. \qquad (4.22)$$

On rearranging, we get

$$t = \frac{1}{\pi \kappa} \left(\frac{T_0}{\text{surface temp. grad.}} \right)^2. \qquad (4.23)$$

This is the formula Kelvin used. It tells us the time that has elapsed since the Earth first became solid and continued to cool till the surface temperature gradient took a particular value. In this case 1°F per 50 ft in depth. Which, with $T_0 = 7000$°F, gives the cooling time as about 98 million years. Allowing for some uncertainty in the initial temperature T_0 and the diffusivity κ, Kelvin put an estimate of the geological age of the Earth as between 20 million and 400 million years.

Bibliography

The three main papers by Kelvin on the Sun and the Earth are the following:

Thomson, W. 1854. On the Mechanical Energies of the Solar System. *Edin. Roy. Soc. Trans.*; *Math. Phys. Papers II* **LXVI**, 1–25.

Thomson, W. 1862. On the Age of the Sun's Heat. *Macmillan's Magazine V*, 288–393.

Thomson, W. 1862. On the Secular Cooling of the Earth. *Edin. Roy. Soc. Trans.*; *Math. Phys. Papers III* **XCIV**, 295–311.

The biographies of Kelvin that I am familiar with are the two-volume set by Silvanus P. Thompson, published in 1910, full of personal details; the book by Andrew Gray (a student and later assistant of Kelvin's); the book by Joe D. Burchfield; and the more modern (excellent) book by David Lindley.

Thompson, S.P. 1910. *Life of Lord Kelvin.* Vols I, II. London: Macmillan and Co., Ltd.

Gray, A. 1908. *Lord Kelvin: An Account of His Scientific Work.* London: I.M. Dent and Co.

Burchfield, J.D. 1975. *Lord Kelvin and the Age of the Earth.* Chicago: University of Chicago Press.

Lindley, D. 2004. *Degrees Kelvin: A Tale of Genius, Invention and Tragedy.* Washington D.C.: Joseph Henry Press.

My own goal was not to write another biography of Kelvin but to understand his calculations.

The basic equations for thermodynamics heat flow are derived for example in

Carslaw, H.W. and Jaeger, J.C. 1959. *Conduction of Heat in Solids*. Oxford: Oxford University Press.

Chapter 5

Radioactivity's Revolution

Radioactivity was discovered in 1896. It was the key that opened the door to deeper levels of Nature. The submicroscopic world. Both atomic and nuclear. And to explain it all — new physics. Twentieth-century physics. The unification of space and time. The uncertainty of quantum mechanics. Such was the revolution that radioactivity demanded.

A well-documented story. With a cast of brilliant characters. Many to receive the Nobel accolade, such as Henri Becquerel, Marie and Pierre Curie, Ernest Rutherford, and Frederick Soddy. Didn't happen overnight. Took decades to decipher.

And for us, a new estimate — based on radioactivity — for the age of the Earth. That's where we're heading now.

Mysteries of the underworld

So what were the main steps? From the discovery of radioactivity to a full understanding of what it was all about. Here's but a summary.

1. *The discovery itself*

Accidental. Henri Becquerel. Looking for a natural source of X-rays, a new phenomenon discovered only a few months earlier by Wilhelm Roentgen, who used complicated cathode ray tubes. Expensive. Couldn't buy one off the shelf. Try something simpler. Like shining light on chemical compounds. See if that would *stimulate* them into emitting X-rays. Just like you can cause some minerals to glow when you shine ordinary light on them. That's called phosphorescence. What about X-rays? Can you produce them in a similar way?

After a while, Becquerel happened to try his luck with a uranium compound. Serendipity? Exposed it to the glorious Paris sunshine for a while. Then placed it on a photographic plate that had been wrapped in thick black paper. To absorb any phosphorescent light that the exposure might have generated in the sample. Stuck it in a drawer overnight. In the morning, developed the plate. Mon Dieu! There on the photographic plate was a fuzzy image of his sample. Must have been caused by something that had been able to penetrate through the protective paper. X-rays? Looks promising.

Same effect when repeated. So no fluke.

Then one day — that fateful day — an overcast sky. No bright Parisian sunshine. But Becquerel had his routine. Which he followed. Lo and behold — same result. Fogging of the plate. As before. *Without* the aid of a bright Sun! Conclusion — the Sun had nothing whatsoever to do with whatever it was he had discovered. The uranium compound was emitting something *on its own*. Didn't need to be stimulated. Marie Curie later called it "radioactivity".

2. *It's everywhere*

Other elements were soon found to be radioactive. First thorium. Then two previously unknown elements. Eked out by the Curies from more than four tons of the uranium-bearing ore pitchblende. (Yes, four tons!) Garnered them less than half a gram of each new element. Polonium (named by Marie Curie after her native land) and radium. And radium was an incredible million times more radioactive than uranium! Obviously very powerful. Immediately caught the attention of medical charlatans who offered it as a "cure" for all sorts of ailments. Eventually causing much distress. And sometimes death.

Marie Curie herself became a victim of radioactivity. Even the lab notebooks she left behind after her death in 1934 were highly contaminated. Now kept under lock and key — along with her radioactive cookbook — in a lead deposit box in Paris.

In fact, we now know that most of the elements in the Periodic Table have radioactive components. Scattered all over the surface of the Earth (including the oceans) in trace amounts. Even in that last cup of coffee you had! No way of escaping it. Radioactivity is everywhere.

3. α, β, and γ rays

So what exactly was fogging up Becquerel's photographic plates? Not X-rays. Something else. More complicated. In fact, three new kinds of rays were identified. Label them α, β, and γ. All with energies much greater than X-rays. By a factor of a thousand or more! Clearly a new phenomenon. Belonging to a different world.

The three kinds of rays could be distinguished from one another in several ways. Such as by their ability to penetrate. α's could typically pass through a few centimeters of air. Or a sheet of paper. β's — more penetrating. Could easily get through several sheets of paper. Even through a sheet of aluminum foil. γ's were the most penetrating of all. Could get through several centimeters of lead!

The rays could also be distinguished by the electric charges they carried — indicated by their response to a magnetic field. Both the α's and β's were deflected, though in opposite directions. In the convention introduced originally by Benjamin Franklin more than a hundred years earlier, it meant that the α's were positive, the β's negative. γ's on the other hand went undeflected. They were electrically neutral.

But what exactly were these rays?

First the α's. Turned out to be doubly charged atoms of *helium*. An element that had been discovered not long before in 1868. Not on Earth, but in the Sun. From a detailed analysis of its light. Eventually found on planet Earth in 1895. In uranium ores. Coincidence? Not really. Put two and two together. The source of the helium in uranium ore was most likely the α's produced by the radioactivity of the uranium it contained.

The β's were also quickly identified. Seen before as the tiny carriers of electric current in cathode ray tubes. Discovered in 1897 by

J.J. Thomson. Given the name *electrons*. Same no matter what kind of hot filament was used to generate the current. Which suggested to Thomson that electrons were something really special — *constituents* of *all* elements. Brilliant insight, especially considering the fact that atoms themselves were only finally being accepted as *real* at the end of the 19th century! In Thomson's view, the negative electrons were most likely embedded in some kind of positive matter (the atom as a whole being electrically neutral). Looked to him very much like a very English "plum pudding".

But Thomson had more to suggest. Maybe the number of electrons that an atom had, reflected the position of the corresponding element in the Periodic Table. One electron for the first element, hydrogen. Two electrons for the next element, helium. Three for lithium. And so on. A pattern. Each element identifiable by the number of electrons in its atoms.

Modern atomic physics and chemistry beginning to emerge right here! (The chemical properties of an element are determined entirely by its electrons.)

The α's and β's were apparently "particles". What about the γ's? Electrically neutral. Behaved more like very energetic X-rays. Continued to be called γ-rays.

So what exactly was happening in radioactivity? Were the atoms themselves "unstable"? Somehow "decaying"? Ejecting α's, β's, and/or γ's in the process? With very high energies. And what was the source of all this energy? A big puzzle ... needed an Einstein to find the answer. Literally!

4. Death knell for Kelvin's Earth

One thing for sure — radioactivity was the final *coup de grâce* for Kelvin's age of the Earth. Even before radioactivity was fully understood.

Why? Because, in 1903, a remarkable fact was discovered about radium — when it decayed, it released *heat*. A lot of it. Enough, it was quoted, to raise the temperature of its own weight of water from freezing to boiling in just one hour! (For this calculation — which relies on Einstein's famous equation $E = mc^2$ — see the Appendix.)

Kelvin had always assumed that there was no source of heat within the Earth. To him, the Earth was just a big hot, solid object, gradually cooling down, dissipating its "primitive" heat into space. Following the laws of physics he knew about.

Radioactivity proved Kelvin wrong. For radioactive materials within the Earth generate heat when they decay. A source of heat Kelvin had not included. Totally undermined his calculations.

Great rejoicing all around by the geologists and evolutionists. The old enemy defeated. At last. By physics — new physics, that is. Radioactivity, and all that implied.

5. *The law of radioactivity*

The *activity* of a radioactive sample (corresponding to how many decays — by whatever means — that take place per unit time) is found to change. It decreases with time.

Fortunately, there's a simple law that describes this property of radioactivity. It's this:

> For any radioactive sample, the number of decays that take place per unit of time is proportional to the (diminishing) amount of radioactive material present in the sample.

Sounds simple. In mathematical terms, with $P(t)$ being the number of "parent" radioactive atoms in the sample at time t, it can be written as

$$\text{activity} = \lambda P(t)$$

where λ is the constant of proportionality. It's got a name — the *decay constant*.

Mathematicians know how to solve this equation, once the activity has been expressed as the rate — represented in mathematics by the derivative $dP(t)/dt$ — at which the number of parent atoms changes with time. Its solution is

$$P(t) = P_o \exp(-\lambda t), \tag{5.1}$$

where P_o is the original number of radioactive atoms in the sample and exp is the exponential function.

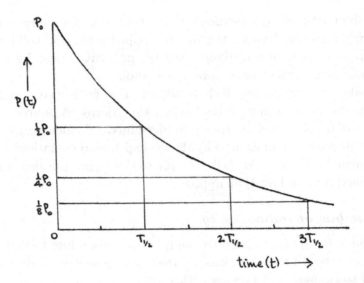

Figure 1:   The exponential decrease with time of the number of radioactive atoms remaining in a sample.

This tells us specifically how the number of unstable atoms remaining in the sample *decreases* with time. It's illustrated in Figure 1.

Fortunately, the decay constant λ is related to another, more physical, quantity. It's called the *half-life*. We'll denote it by $T_{1/2}$. The *half-life* is defined as how long it takes for the number of radioactive atoms in a sample to diminish to half of what it was initially (see Figure 1), that is, the time it takes for $P(t)$ to decrease from P_o to $\frac{1}{2}P_o$. Substituting this into Eq. (5.1) above gives

$$\frac{1}{2}P_o = P_o \exp(-\lambda T_{1/2}).$$

Rearrange this and take the logarithm of both sides. You get

$$\lambda = (\ln 2)/T_{1/2}, \tag{5.2}$$

where $\ln 2$ is the Naperian logarithm (that is, the logarithm to the base e) of 2. Its value is 0.693.

What happens to the "parent" when it decays? It becomes a "daughter". (Who decided on its gender, I don't know!) Let's assume for the sake of discussion that the daughter is stable. Then the number of daughter atoms $D(t)$ formed in the time t by the decay of the parents is given by

$$D(t) = P_o - P(t)$$

$$= P(t) \exp(\lambda t) - P(t)$$

$$= [\exp(\lambda t) - 1]P(t), \qquad (5.3)$$

where I've eliminated P_o (usually not known) in favor of $P(t)$ (which can be measured). Equation (5.3) gives the number of daughter atoms generated in time t by the radioactive decay of the parent atoms. A graph of $D(t)$ against time t is shown in Figure 2.

As we can see, the amount of the daughter *grows* exponentially with time, in step with the exponential decline of the amount of the parent.

Deep down, radioactivity is a statistical process, governed by the probabilistic rules of another great revelation of the 20th century — quantum mechanics. You can't tell exactly when an individ-

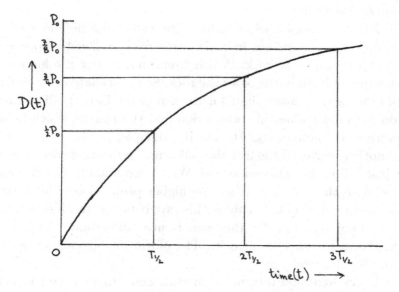

Figure 2: The exponential growth of stable daughter atoms from radioactive parents.

ual parent atom will decay. Could be any time. The best you can do is to consider a large number of them and determine how long it takes, on average, for half of them to decay. That is what the half-life represents.

Half-lives can vary tremendously, from microseconds to billions of years. Radium for example has a half-life of 1600 years. What's really important is that each radioactive material has its own *unique* half-life. That's how the early workers in the field were able to separate out one radioactive species from another. And there were lots of them. In addition, half-lives were found to be unaffected by temperature or pressure. Which makes them particularly suited for determining the age of the Earth.

6. *Early attempts*

The first attempts to measure the age of the Earth by means of radioactivity were pretty crude. At least by modern standards. The reason? Simply — in the early 1900s, there just wasn't enough known about radioactivity.

Take, for example, what came to be called the helium method. Assumed that the helium in rocks came from radioactive processes (α-decays) within the rock. Which would mean that the higher the proportion of helium, the older the rock. So to estimate the age of the rock (and hence a lower limit for the age of the Earth), all you need to do is to determine this proportion and the rate at which helium is produced. Sounds easy. At least in principle.

Another way used the fact that all uranium-bearing rocks contain not just helium but also some lead. Which presumably was also associated with the uranium. If so, the higher proportion of lead would indicate an older rock. Combine this proportion with an estimate of the accumulation rate for the lead. Hence, an estimate for the age of the rock. Do for lots of rocks. The oldest one gives a lower limit on the Earth's age.

All very logical. Problem — in each case, there were too many unknowns or poorly known quantities. Such as — how much helium or lead was there to begin with. Before they were supplemented by radioactive decays. And some of that helium may have escaped, lost. So lots of questions about the conclusions. Still, one thing common to all these results — they pointed to an age of a billion years or more. Yes, billion. Not million.

It was a start. To get further, much more had to be learned about radioactivity. It was a lot more subtle than anyone ever imagined.

7. *Alchemy*

Means one element changes into another. What many scientists of old had tried to achieve. Especially if the end product was silver or gold (read money). Never succeeded.

But alchemy is exactly what happens in α- and β-radioactivity. One element changes, or *transmutes*, into another!

A staggering discovery. When the parent changes — or "decays" — into its daughter, that daughter is a different element.

And just as amazing, as often as not, this daughter itself is radioactive. With a different, unique half-life. Decaying into yet another element. And so on. A whole cascade of sequential decays — a *decay chain* of related radioactive elements.

All of which needed to be sorted out. Requiring very intricate analyses. Much of the early work done by Ernest Rutherford and colleagues. No surprise he was awarded a Nobel Prize in 1908 for his many insights and contributions. But not in Physics — it was in Chemistry. As Rutherford quipped later — it was the quickest transmutation of a physicist into a chemist he had ever known!

8. *An atom's internal structure*

An atom is not really like a plum pudding. Proved by experiment. It's mostly empty space. With a nucleus.

1911 was the momentous year when this was discovered. α-particles from radium were fired at a sheet of gold foil. Simple enough. Most of the α's went sailing through. Just what you'd expect from a mushy plum pudding picture — nothing to turn the α's around. Problem ... incredibly, a few of the α's did bounce back! Atoms couldn't be plum puddings.

So, what could explain all this? The atom had to look very different from what people had previously thought.

For starters, all the atom's positive charge had to be concentrated. At the center. To ensure the strongest possible repulsion toward an incoming positive α. To turn it around.

In addition, the atom's mass had to be concentrated at the center. That too would facilitate the bounce back.

Conclusion — the atom had to have a small core where its positive electric charge and mass were concentrated. Call this core the *nucleus*. And where were Thomson's negative electrons? They must be buzzing around the outside, at a distance.

Quite a revelation. A totally new picture of the atom.

But there was more. Two years later. That's when it was realized that these electrons weren't just buzzing around the nucleus randomly. They were marching to the tune of a new drummer. In *distinct orbits*. Determined by new physics — *quantum physics*. Explained, for example, as no model had ever been able to do before, the uniqueness of the light that atoms emit when they get "excited" — such as when they are put in flames, or firecrackers, or stars! Not a continuous rainbow of colors. Only specific, discrete wavelengths. What is called a *line spectrum*. Such as that of the hydrogen atom shown in Figure 3 (a nanometer is 10^{-9} meter). Each element has its own unique line spectrum. Like a fingerprint.

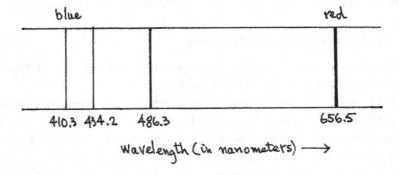

Figure 3: The line spectrum of atomic hydrogen.

And incredible advance. Certainly deserved a Nobel Prize for the young Niels Bohr who thought up the idea. Trouble was, Bohr's model of the atom and how it worked depended crucially on another instance of Einstein's insight — this time about the nature of light. Had two simultaneous characteristics, Einstein said, not just one. Wave and particle. Which were related. Explained a lot of data. So if Bohr was to be awarded a Nobel Prize, didn't Einstein deserve one too? The Nobel Committee came up with an ingenious way of awarding Prizes to both men. It was 1922. By coincidence, the Nobel Prize in Physics the year before had not been awarded. So it was decided to *retrospectively* award the 1921 Prize to Einstein — in 1922. The same year as Bohr.

No such luck for Rutherford though. The father of the nuclear model of the atom itself. Passed over. Presumably not worthy enough. Like Einstein's other great creation "The Special Theory of Relativity". Nor his even greater "General Theory of Relativity". Never recognized. Just not good enough.

9. *Isotopia*

Still can't calculate the age of the Earth reliably. More complications. Life's not easy. Nor is physics. Apparently, the atoms of elements are not all identical. Like ice cream, they can come in different flavors. Isotopes, they're called.

This surprising property of atoms was suspected early on, once the existence of decay chains was realized. More than one kind of uranium, for example. And of lead, the stable end product in the uranium decay chain. Frederick Soddy suggested there might in fact be as many as forty radioactive links in the uranium decay chain. Even though there were only eleven elements from uranium to lead.

How could that be? Suggestion — each element has more than one form. Hence, the idea of isotopes.

Trouble then is — how do you separate the various isotopes of an element from one another? Can't be done by chemistry since the isotopes of an element have exactly the same number of orbiting electrons. So their chemical properties are identical.

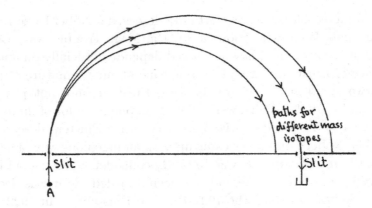

Figure 4:   Deflection of different isotopes of an element by a magnetic field.

However, they can be distinguished by their masses. For it turns out that the different isotopes of the same element have different masses. Which means that, when they are propelled through a magnetic field, the different isotopes will be deflected by slightly different amounts. Led to the construction of powerful new devices called *mass spectrometers*. To tease out the intricacies of isotopes.

A schematic diagram illustrating the principles of a mass spectrometer is shown in Figure 4.

In Figure 4, a beam of charged atoms (known as "ions") of an element is formed by accelerating the ions from A through a narrow slit. Its path is subsequently bent by a magnetic field. (In Fig. 4, the magnetic field is perpendicular to the page.) The amount of deflection however depends on the mass of the ion. Which results in the various isotopes in the sample being separated, as shown. The individual beams can be guided in turn to pass through a second slit, and their masses and relative abundances measured.

Quite a revelation. All the elements had isotopes!

10. *Protons and neutrons*

What clues did the masses of the elements and their isotopes provide? In particular, were there any patterns? (Which is always one of the first questions a physicist asks.)

Start small. Such as with the simplest of the elements. The ones at the beginning of the Periodic Table.

Hydrogen. What do we know about it? Has one electron in orbit. Negatively charged. Determines the chemical properties of hydrogen. Its nucleus carries most of the mass of the whole atom. And is positively charged. Presumably has the simplest nucleus of all. Call it a *proton.* (Its mass is in fact 1836 × the mass of an electron.)

Now the next element — helium. What insight can helium give us? Its atoms have two electrons in orbit. So there must be two protons in its nucleus to balance off the electric charge. If this were the case, a helium atom should be about twice the mass of a hydrogen atom. But it isn't. It has about *four* times the mass! How come? Something missing?

What about the third element — lithium? Three electrons in orbit. So expect three protons in the nucleus. Which would make the mass of a lithium atom about three times that of a hydrogen atom. Again, it's not! It has about *seven* times the mass. Definitely something's missing.

Rutherford — that man again — had a suggestion. It was this. Perhaps there's another constituent in nuclei besides the proton. With about the same mass. Carrying no electric charge (so as not to upset the electric balance between the orbiting electrons and the protons). Given the name *neutron.*

And how can this explain the fact that say helium atoms have about four times the mass of a hydrogen atom? That's because, according to Rutherford, in their nuclei, helium atoms have not only two protons but also two neutrons. A total of four nuclear constituents or *nucleons* as they are called ("nucleon" being the generic name for either a proton or a neutron).

What about lithium? Its nucleus must contain not just three protons but also four neutrons. A total of seven nucleons. Making lithium atoms have a mass about seven times the mass of hydrogen atoms. Which is in agreement with observation.

You get the idea. The nucleus of the atom of an element has a unique number of protons corresponding exactly to the number of orbiting electrons (thus making the atom electrically neutral). Which fixes that element's place in the Periodic Table (and its chemistry). In addition, there are neutrons which make up the atom's remaining mass.

11. *Variations on a theme*

The number of protons in the atoms of a particular element is fixed. What about the number of neutrons? Fixed, or could there be several possibilities? Answer — the number of neutrons can vary. Which is the origin of isotopes.

Let's go back to hydrogen. So far, we've considered its nucleus as containing just one proton. Could there be another kind of hydrogen which had say one neutron in its nucleus besides its defining single proton? It would have the same chemical properties as hydrogen atoms with just one proton. But about twice the mass. "Heavy" hydrogen! Does it exist?

The answer is yes. Though it took a long time to find. Discovered in 1931 by Harold Urey. Identified it by the unique pattern of its line spectrum — predicted by Bohr. Gave it the name *deuterium*. A second kind of hydrogen. Not much of it around. But it is there — about 0.0156% of all the hydrogen in the oceans is of this kind. Used in the 1940s to make nuclear reactors. As "heavy" water. Also the key isotope of hydrogen in the nuclear processes that keep our Sun shining. And us alive! Immediate Nobel recognition for Urey. In Chemistry, 1934. "The first prize I ever won!"

In fact, all the elements have isotopes. There's no unique recipe for the nuclei of atoms that belong to a particular element. They all have the same number of protons. But the number of neutrons can vary. That's what distinguishes one isotope of an element from another. Hydrogen even has a third isotope — tritium. Has two neutrons in the nucleus besides the proton. But unlike deuterium, tritium is *unstable*, that is — it's radioactive. Half-life 12.33 years. The three isotopes are sometimes referred to as hydrogen-1, hydrogen-2, and hydrogen-3. The numbers after the element's name indicate the number of *nucleons* in the nucleus.

Another well-known example of isotopes is carbon-12 and carbon-14. You've probably heard of them. Carbon is the sixth

element in the Periodic Table. Which means that all of its atoms
have six orbiting electrons and six protons in their nuclei. Carbon-
12 has in addition six neutrons in its nucleus. Carbon-14 has eight.
Carbon-12 is stable, but carbon-14 is unstable — radioactive with a
half-life of 5,730 years. Which makes it ideal for radio-dating organic
materials (such as wood and bone) that are several thousand years
old. Revolutionized archaeology.

Uranium likewise has several isotopes, the most important of
which, for our purposes of determining the age of the Earth, are
uranium-238 and uranium-235. Uranium is element 92. That is, all
of its isotopes have 92 protons in their nuclei. Uranium-238 has in
addition 146 neutrons. That's quite a few! Uranium-235 has 143 neutrons. The biggest and most massive nuclei that exist in Nature.

With that number of neutrons, it's not surprising to learn that
both of these uranium isotopes are radioactive. With incredibly long
half-lives: 4.468 billion years and 0.704 billion years, respectively.
Which makes them perfect for determining the age — scientific, that
is — of the Earth.

12. *Neutrons — fact or fiction?*

Neutrons are certainly convenient for explaining isotopes. For
counting. First suggested by Rutherford in 1920. But were they real,
physical entities?

A familiar question throughout the history of physics. Like atoms
in the 19th century. John Dalton in the early 1800s suggested that
all matter is made of atoms. Combined in definite ratios to form
"compound atoms" (read molecules). Explained a lot of chemistry.
Good for counting. But real? It took almost a hundred years before
they were finally accepted.

And more recently. The same thing with quarks in the 1960s. The
supposed building blocks of protons and neutrons themselves. Had
peculiar properties. But they did give a clear explanation for the
spectrum of "excited" nucleonic states. Beautifully! Only a handful
of outcasts believed they were real. Too far-fetched for most. But
times do change — I don't know anybody nowadays who doesn't
"believe" in quarks!

So what about neutrons. Real or not? Best way to answer this question — find one! Which is exactly what James Chadwick did in 1932. Bombarded beryllium with α-particles from radioactive polonium. Produced neutral "rays". Were they γ-rays? No, they had mass, essentially the same mass as a proton. Had to be neutrons, Rutherford's second ingredient for atomic nuclei.

Chadwick had been hunting for neutrons for years, under Rutherford's direction. He had originally intended to major in mathematics at university. Joined the wrong queue at registration. Ended up doing physics. As a graduate student, he went to work in Berlin with Hans Geiger (of Geiger counter fame ... developed while working in Rutherford's group in England). Arrested there in 1914 at the beginning of the First World War. A foreign alien. Had to waste four long years in a German internment camp before being finally released at the end of the War. But triumph in 1932. Chadwick proved that neutrons really were real! As real as protons. Confirmed Rutherford's nuclear model of the atom. A Nobel Prize for Chadwick in 1934. (None for the great man himself.)

13. *More on α, β, and γ decays*

A few more words about α, β, and γ decays.

First, α-decay. Exactly what is happening in the nucleus? It involves a tight cluster of two protons and two neutrons — that is, an α-particle — which manages to penetrate through the barrier that normally constrains all the nucleons to stay within the nucleus. The half-life of the corresponding decay reflects the *probability* of this unusual event happening. Can all be worked out mathematically, from the great fundamental equation of quantum mechanics — the Schrodinger equation. Discovered by Erwin Schrodinger while on vacation in the Austrian Alps over Christmas 1925. With his girlfriend from Vienna. Inspired. (His wife Anny had decided to stay home in Zurich.) Schrodinger's equation allows us to *calculate* this probability, and hence α-decay.

β-decay is very different. Comes about when a neutron *inside the nucleus* changes into a proton, ejecting an electron in the process.

The corresponding process may be written as

$$n \to p + e^- + \overline{\nu_e},$$

where n and p represent the neutron and proton, respectively, and e^- the ejected electron. Note that this is a *newly created* electron (with lots of energy), not one of Bohr's orbiting electrons.

Note also that there's another particle created in β-decay, denoted by $\overline{\nu_e}$. It's called the electron antineutrino. First proposed in 1930 to ensure, for example, that the basic law of energy conservation was not violated in this decay. Very elusive. Not actually detected experimentally till 1956. Has a partner called the electron neutrino. Lots of them are created in the Sun by the nuclear processes that generate the Sun's heat. On which we depend. When I say "lots", I mean "lots". Sixty-five billion of them are whizzing through your body every second. You're joking, you say. No. It's true. Astrophysicists know how the Sun works!

γ-rays are totally different again. They're emitted when a daughter nucleus is formed with excess energy. It gets rid of this excess energy by emitting one or more γ-rays. Which is what happens when, for example, cobalt-60 decays. That's the isotope that's often used in the medical world to treat some types of cancer. The cobalt-60 β-decays to nickel-60. Which is formed in what's called an "excited state". When it de-excites, the nickel-60 emits γ-rays. Which are the killer rays used in medical therapy.

14. *Notation for isotopes*

There's an alternative way of specifying isotopes than the one we've been using so far. Such as the three hydrogen isotopes hydrogen-1, hydrogen-2, and hydrogen-3. Or the carbon isotopes carbon-12 and carbon-14.

For the three hydrogen isotopes, it's the following:

$$\mathrm{^1_1H}, \quad \mathrm{^2_1H}, \quad \text{and} \quad \mathrm{^3_1H}.$$

You can easily recognize what the various bits and pieces stand for. The chemical symbol H for hydrogen. The common suffix 1 corresponds to the fact that hydrogen is the first element in the Period

Table; it's also the number of protons in the nucleus — the same for all its isotopes. The superscript is the total number of nucleons (that is, protons plus neutrons) in the nucleus. This is the number that is different for the various isotopes.

Likewise for carbon. The two isotopes carbon-12 and carbon-14 can be written as

$$^{12}_{6}C \quad \text{and} \quad ^{14}_{6}C,$$

respectively.

And in the determination of the age of the Earth, we'll be interested in the uranium isotopes uranium-238 and uranium-235, which we can now denote by

$$^{238}_{92}U \quad \text{and} \quad ^{235}_{92}U.$$

Why bother with this new notation? As we'll see, it's convenient for keeping track of the constituents of the nuclei in radioactive processes.

15. *The uranium decay chains*

According to science, the Earth is old. Really old. Billions of years. Not millions as Kelvin had wanted. Billions is what the early estimates of the Earth's age indicated from radioactivity. But it needed to be done more accurately.

Not easy trying to measure such huge lengths of time. Ordinary clocks have time units of minutes and hours. Pretty useless here. What we need are clocks whose basic time units are billions of years. Sounds crazy. Are there any such clocks? Actually, yes. Such as uranium-238. Its basic time unit is its half-life. Which is 4.468 billion years. Or uranium-235 whose half-life is 0.704 billion years. Unaffected by wind or weather. Or in solid rock or molten lava. Reliable. Both of them ideal clocks for this purpose.

So what do we know about them? Took a long time to figure out. Fortunately, only a few decades!

In both cases, their radioactive stories are complicated. Each involves a whole cascade of decays, transmuting step by step through a long series of other elements. Eventually ending in lead.

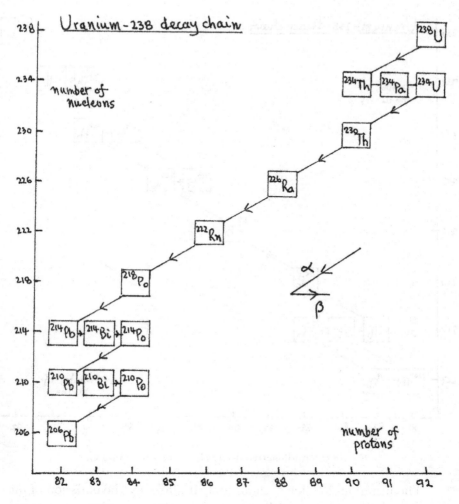

Figure 5: Radioactive decay chain for uranium-238.

The full decay chains are shown in Figures 5 and 6. The vertical axes correspond to the number of *nucleons* in the nucleus, the horizontal axes give the number of *protons* (and hence the particular element involved).

Pretty impressive. You can see why it took so long to figure out.

Let's take a quick look at each decay chain before going on to see how they are used to determine the age of the Earth.

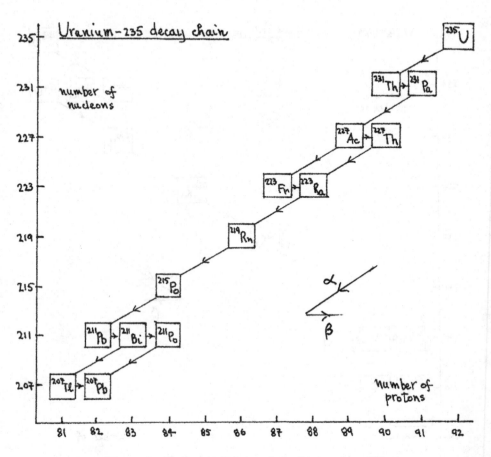

Figure 6: Radioactive decay chain for uranium-235.

The uranium-238 decay chain first. It starts by the emission of an α-particle (helium-4), which removes two protons and two neutrons, a total of four nucleons. Summarized as

$$^{238}_{92}\text{U} \rightarrow {}^{234}_{90}\text{Th} + {}^{4}_{2}\text{He}.$$

The uranium-238 transmutes into thorium-234. Which itself is unstable and β-decays into an isotope of proto-actinium (an element few people have heard of). Which in turn β-decays into uranium-234; which α-decays to ... all the way down to the stable isotope lead-206. The end of the decay chain. After a total of eight α-particles and six β-particles being ejected.

The uranium-235 decay chain is somewhat similar. The first transmutation again involves an α-decay, this time to the isotope thorium-231:

$$^{235}_{92}\text{U} \rightarrow {}^{231}_{90}\text{Th} + {}^{4}_{2}\text{He}.$$

Just the first of many subsequent transitions. Ultimately ending in a different lead isotope lead-207. After a total of seven α-decays and four β-decays.

The chemical symbol for lead looks a bit strange. Pb. Where does that come from? From the Latin word for lead: plumbum. Not as uncommon in the English language as you might think. How about "plumbers" — the artisans of old who fixed lead pipes and church roofs. Or a "plumb line" — a string with a lead weight attached, used to indicate the vertical? Occasionally, the stock market "plummets". And of course you would always like to act with great aplomb. That plumbum is everywhere!

Billions. Not Millions. Or Thousands

The master equation

Now for the final steps. It has taken us a long time to get all the bits and pieces in place. Not surprising. An old world had to be replaced by a new one. Which required time for it to be explored and understood before the new knowledge could be applied properly to determine the age of the Earth. It all came together in the late 1940s, early 1950s. Based primarily on the uranium–lead decay chains. Here's the gist ...

First, a quick look again at the uranium decay chains. Both of them. Plotted out in Figures 5 and 6. They have much in common. One particularly striking feature: the half-lives involved. For both chains, the first decays (uranium-238 to thorium-232, and uranium-235 to thorium-231) have exceptionally long half-lives: 4.468 billion years and 0.704 billion years, respectively. But after that, all the succeeding decays are found to be comparatively rapid. With end-products lead-206 and lead-207.

This surprising characteristic of these two uranium decay chains can be used to our great advantage to simplify our calculations.

The reason is this: it's a good approximation to think of uranium-238 as a parent that transmutes essentially to lead-206 with a half-life of 4.468 billion years. Likewise for uranium-235: it transmutes essentially into lead-207 with a half-life of 0.704 billion years. To all intents and purposes, we can ignore the brief intermediate transitions. Makes the equations a lot easier! (Something you will probably approve of!)

A comment about the lead that's found in rock samples: there are in fact two sources for this lead. There is of course the lead that is produced, as we have just described, from the decay of uranium in the sample. That's usually referred to as the "radiogenic" component. It increases with time as the uranium continues to decay.

But there's also the possibility that there was some lead-206 and lead-207 in the rock to begin with, when it first formed out of its molten state. That's called the "primeval" or "primordial" component. We must not forget to include this in our calculations.

All of which allows us to write down an equation for each of the lead isotopes, assuming that the rock was formed T years ago:

$$^{206}\text{Pb} = [\exp(\lambda_{238}T) - 1]\,^{238}\text{U} + {}^{206}\text{Pb}_0. \tag{5.4}$$

$$^{207}\text{Pb} = [\exp(\lambda_{235}T) - 1]\,^{235}\text{U} + {}^{207}\text{Pb}_0. \tag{5.5}$$

Each of the right-hand sides is the sum of the radiogenic component (from Eq. (5.3)) and the primeval component, which we denote by $^{206}\text{Pb}_0$ and $^{207}\text{Pb}_0$.

Some comments about the notation in these equations. As is conventional, the isotopic *symbols* are meant to represent the corresponding *amounts* of the isotopes. ^{206}Pb, ^{207}Pb, ^{238}U, and ^{235}U represent the amounts of the various isotopes at the present time, that is, T years after the rock formed. $^{206}\text{Pb}_0$ and $^{207}\text{Pb}_0$ are the primeval amounts of these isotopes. λ_{238} and λ_{235} are the (known) decay constants of the two uranium isotopes.

Next, an improved way of doing things. Instead of working directly with the individual isotopes, it's much better to measure them with respect to some other isotope, one that remains constant over time. Use it as a reference. So work with ratios, rather than direct amounts.

And which isotope can we use for this? Lead-204. It's another isotope of lead. But it's stable; doesn't decay. And it's not the end product of any decay chain. So it's a constant. Doesn't change with time. Ideal for our purposes.

So now we are going to rewrite Eqs. (5.4) and (5.5) above as ratios:

$$\left(\frac{^{206}\text{Pb}}{^{204}\text{Pb}}\right) = [\exp(\lambda_{238}T) - 1]\left(\frac{^{238}\text{U}}{^{204}\text{Pb}}\right) + \left(\frac{^{206}\text{Pb}_o}{^{204}\text{Pb}}\right) \quad (5.6)$$

$$\left(\frac{^{207}\text{Pb}}{^{204}\text{Pb}}\right) = [\exp(\lambda_{235}T) - 1]\left(\frac{^{235}\text{U}}{^{204}\text{Pb}}\right) + \left(\frac{^{207}\text{Pb}_o}{^{204}\text{Pb}}\right). \quad (5.7)$$

Note that since lead-204 is constant, ^{204}Pb and $^{204}\text{Pb}_o$ are the same.

One last step. Which makes use of one additional piece of information. Namely, that at the present time, the two uranium isotopes ^{235}U and ^{238}U are always found in ores in the same ratio. Everywhere. And it's measured to be

$$\left(\frac{^{235}\text{U}}{^{238}\text{U}}\right) = \frac{1}{137.88}. \quad (5.8)$$

Combine this now with Eqs. (5.6) and (5.7). It leads to our master equation. The algebra is a bit tricky. But here's what to do — move the primeval contributions on the right-hand sides over to the left-hand sides, then divide the two equations. Which gives

$$\frac{\left[\left(\frac{^{207}\text{Pb}}{^{204}\text{Pb}}\right) - \left(\frac{^{207}\text{Pb}_o}{^{204}\text{Pb}}\right)\right]}{\left[\left(\frac{^{206}\text{Pb}}{^{204}\text{Pb}}\right) - \left(\frac{^{206}\text{Pb}_o}{^{204}\text{Pb}}\right)\right]} = \frac{1}{137.88}\left[\frac{\exp(\lambda_{235}T) - 1}{\exp(\lambda_{238}T) - 1}\right]. \quad (5.9)$$

Looks pretty ghastly. But it's really just the complicated notation that's throwing us off. Believe it or not, Eq. (5.9) is a *straight line* in disguise!

To see this, let's change the notation: set

$$y = \left(\frac{^{207}\text{Pb}}{^{204}\text{Pb}}\right), \quad y_o = \left(\frac{^{207}\text{Pb}_o}{^{204}\text{Pb}}\right). \quad (5.10)$$

$$x = \left(\frac{^{206}\text{Pb}}{^{204}\text{Pb}}\right), \quad x_o = \left(\frac{^{206}\text{Pb}_o}{^{204}\text{Pb}}\right). \quad (5.11)$$

$$m = \frac{1}{137.88} \left[\frac{\exp(\lambda_{235}T) - 1}{\exp(\lambda_{238}T) - 1} \right]. \tag{5.12}$$

Then the complicated-looking Eq. (5.9) becomes

$$\frac{(y - y_0)}{(x - x_0)} = m,$$

that is

$$(y - y_0) = m(x - x_0) \tag{5.13}$$

or simply

$$y = mx + k, \tag{5.14}$$

where $k = (y_0 - mx_0)$. Okay, you recognize it now. A straight line with slope m and intercept k. In the (x, y) coordinate plane. This is our master equation.

It's the slope m that's the crux of this whole procedure. It depends only on T, the age of the rock. Repeat — *only* on T. So determine the slope m and you've got T.

Isochrons

Let's take a closer look now at Eq. (5.14).

Here, the x-axis corresponds to the *present* proportion (relative to lead-204) of the isotope lead-206. It is measured for each of our various rock samples. You might think that, since the samples are all taken from the same rock, they should all have the same value of $^{206}\text{Pb}/^{204}\text{Pb}$. That was certainly the situation when the rock was molten; everything homogenized. But that can change as the rock cools down and solidifies. As quantum chemists tell us, crystallization is a very subtle process, very dependent, for example, on the temperature. So as the temperature of the rock falls, first one kind of crystal will form (thereby changing the element composition in the remaining rock), then other kinds. The net result — different samples, taken T years later from the same solid rock, can have different values of $^{206}\text{Pb}/^{204}\text{Pb}$. In other words, the different samples have different x-coordinates.

Likewise, for the present proportions ^{207}Pb/^{204}Pb in these samples. They can all be different. Which means that their y-coordinates can be different.

Now put the x- and the y-coordinates of each sample together, forming the coordinates of the point (x, y). Which you can mark on the x-y plane, each sample being represented by a single point.

And where do all these points lie on the plane? Scattered all over? Not so. Remember that, even though the samples have different proportions of the isotopes, they were all formed at the same time T years ago. So their corresponding points on the x-y plane must satisfy Eq. (5.14). That is, they must all lie on the *same* straight line. Whose slope is m. Which depends solely on T. So by determining the slope, you can immediately deduce the common age of all the rock samples.

The general idea is illustrated in Figure 7.

Equation (5.14) in fact defines a whole family of straight lines which differ from one another by the value of their slope m. Called *isochrons*. Meaning "equal time". For all the points on any one isochron correspond to rock samples that all have the same age. And from Eq. (5.12), the greater the age T, the greater the slope m.

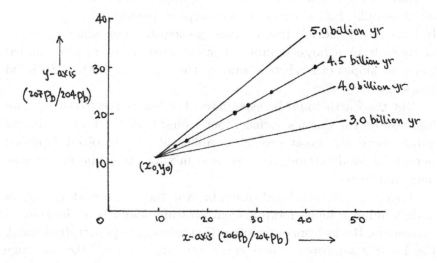

Figure 7: Isochrons.

Clearly, for the rock illustrated, its data points lie along the isochron that corresponds to an age of 4.5 billion years.

An aside — notice that all the isochrons pass through the one point (x_o, y_o). How come? Chemistry. For crystallization does not differentiate between the isotopes of an element. So, all the samples start off with the same values for $^{206}Pb_o/^{204}Pb$. And from Eq. (5.13): when $x = x_o$, then $y = y_o$ whatever the value of the slope m.

The age of the Earth

There is of course no way of determining the age of the Earth by analyzing rock samples as described above. Too bad. All you can do is determine the age of the rock from which the samples are taken. Which is *not* the age of the Earth. But what you can get out of this, by examining all the rocks you can lay your hands on, is the age of the *oldest* rock on the surface of the Earth. That gives you a *minimum* age of the Earth.

So where's the oldest rock on the surface of the Earth? Apparently, there are old rocks in lots of places — often with exotic names like Isua in southwest Greenland, Navvuagitting on the eastern shores of Hudson Bay, Acasta in northwest Canada, and Pilbara in Western Australia. Following in the footsteps of pioneers such as Arthur Holmes and Friedrich Houtermans, geologists determined the ages of rocks from a large number of sites. Their answers, determined from the slopes of isochrons, were in the region of 3.0 to 3.8 billion years ago.

But the Earth could be even older. For before the oldest rock we find now on the Earth's surface, there could have been a time during which there was great geological upheaval that involved repeated periods of solidification, melting, and mixing. And who knows how long that lasted.

Then, in 1953, a breakthrough. No, the eagle-eyed geologists hadn't found a much more ancient stratum somewhere. Instead — *meteorites*. Rocks from outer space that manage to penetrate through the Earth's atmosphere and crash onto its surface. Like the huge monster that is thought to have led to the demise of the dinosaurs

65 million years ago. Or the 300,000-ton rock that slammed into the desert of northern Arizona 50,000 years ago. Formed a crater three quarters of mile wide — Meteor Crater. (Go visit it if you are anywhere nearby!) With lots of fragments scattered around it. Especially in Canyon Diablo nearby. Analysis of one of these fragments in 1953 by Clair Patterson and colleagues led to a new revelation.

Apparently, the meteorite they were investigating contained essentially no uranium. But did contain lead. All three isotopes lead-204, 206, and 207. Since there was no uranium present, it meant that none of the lead was radiogenic. That is — all of it was primeval. Their proportions yielding $^{206}Pb_o/^{204}Pb$ and $^{207}Pb_o/^{204}Pb$. For meteorites.

But, if meteorites were formed as part of the early solar system, orbiting in isolation, too small (unlike the Earth) to undergo any geological change, then these results for $^{206}Pb_o/^{204}Pb$ and $^{207}Pb_o/^{204}Pb$ must be the primeval ratios of lead isotopes for the whole solar system.

Which fixes the point (x_o, y_o) in our isochron diagram Figure 7. The anchor point for all isochrons. The other data — from samples taken from other meteorites that contain some uranium as well as lead — fix the particular isochron. Whose slope indicates how long ago it was since the solar system — including the Earth — first formed: 4.5 ± 0.3 billion years.

Sounds good. But how reliable?

Very reliable! Checked and cross-checked many, many times. Lots of different meteorites. Even rock samples from the Moon!

And not just using uranium–lead decays. For there are many other pairs of isotopes that can be used. Such as the rubidium–strontium decay (^{87}Rb to ^{87}Sr). Or the potassium–argon transition (^{19}K to ^{18}Ar). All consistent. Giving the same answer. By questioning. Measuring. Analyzing. Confirming. It's the nature of science. To determine the science of Nature.

So the final answer for the age of the Earth? 4.543 billion years. Give or take a few million years. That's the science answer.

Appendix: Heat from radioactive decay

In March of 1903, Pierre Curie and his assistant Albert Laborde discovered that radium somehow generates heat. Enough they said to raise the temperature of its own weight of water from freezing to boiling in one hour. That's a lot of heat. Immediately threw into question Kelvin's calculation for the age of the Earth. The last nail in its coffin.

Energy from mass

But where did all this new-found heat energy come from? Not from a chemical process. Not enough energy there. Must be something totally new.

An intractable puzzle. At least for 19th-century physics. Step forward Albert Einstein. Him with the iconic bushy hair. In 1905 (in the days before he developed his bushy hair). Unified space and time. Mind boggling. With lots of consequences. Like length contraction and time dilation. Called the Special Theory of Relativity. Also, in his second paper on the subject, the most famous equation in all of physics:

$$E = mc^2.$$

Who hasn't heard of it? Explains (among other things) where the heat comes from in radioactive decay.

What does Einstein's equation tell us? That mass and energy are related to each other. Not two distinct entities as previously thought. Mass–energy equivalence. It says: with any mass m, you can associate an equivalent amount of energy E, given by the above equation. The quantity c is the speed of light.

Let's see how it applies to radioactivity. In particular, how is heat generated when radium decays?

In terms of isotopes, the decay we are interested in (see Figure 5) is

$$^{226}_{88}\text{Ra} \rightarrow {}^{222}_{86}\text{Rn} + {}^{4}_{2}\text{He},$$

in which radium-226 transmutes into radon-222 with the emission of an α-particle.

Now let's compare the mass of the initial radium-226 atom with the sum of the masses of the decay products. The (measured) masses involved are as follows:

$$^{226}_{88}\text{Ra}: \quad 226.0254\,\text{u}$$
$$^{222}_{86}\text{Rn}: \quad 222.0176\,\text{u}$$
$$^{4}_{2}\text{He}: \quad 4.0026\,\text{u},$$

where u is a standard unit of mass that chemists like to use (it's one-twelfth of the mass of a carbon-12 atom).

Notice that, if you add together the masses of the radon and helium, you get a total mass of 226.0202 u. Which is *less than* the mass of the parent radium. By an amount 0.0052 u.

Where did that missing mass go? Enter Einstein. It appears as energy. The two end products carried it away as kinetic energy. Then through collisions with the neighboring atoms, that kinetic energy was converted into heat energy.

How much energy?

It's the energy that corresponds to the decrease in mass between the parent and the two decay products. That is, to the energy equivalent to a mass difference of 0.0052 u. Which we can be determined by applying $E = mc^2$.

First, the energy equivalent to one mass unit u, which is 1.661×10^{-27} kg. Apply $E = mc^2$, where c is the speed of light 2.998×10^8 m per sec. You'll find the equivalent amount of energy is 1.493×10^{-10} J.

Hence, the energy released in the decay of *one* radium-226 atom is $(0.0052)(1.493 \times 10^{-10}$ J), which is 7.76×10^{-13} J.

Pretty small!

But wait a minute. A big number coming up.

Atoms in one gram of radium-226

What we're really interested in is the heat generated not by just one atom of radium-226, but by say one gram of it. And for this, we need to know how many atoms there are in 1 gram of radium-226.

How do we find that? Call on another sacred number in physics and chemistry. Avogadro's number. Which tells us that in 226 grams of radium-226, there are 6.022×10^{23} atoms. A big number this time! From which it follows that, in 1 gram of radium-226, there are 2.66×10^{21} atoms.

Energy generated in 1 hour

Now in 1 gram of radium-226, not all of the atoms decay at once. In a flash. Rather, it's a long drawn-out process. Governed by quantum mechanics. Encapsulated in the law of radioactive decay. Which tells us that

(number of decays per hour in a sample)

$= \lambda \times$ (number of unstable atoms in the sample),

where λ is the decay constant — for radium-226 in the present case. Related to its half-life by $\lambda = 0.693/(\text{half-life})$. The half-life of radium-226 is 1600 years, or 1.40×10^{7} hr. So $\lambda = 0.495 \times 10^{-7}$ hr^{-1}.

Thus, the heat generated by 1 gram of radium-226 in 1 hour is given by

(energy generated in one decay)(number of decays per hour)

$= (7.76 \times 10^{-13} \text{ J})(0.495 \times 10^{-7} \text{ hr}^{-1})(2.66 \times 10^{21})$

$= 102 \text{ J/hr.}$

Heating water

And how does this compare with the amount of heat you need to raise the temperature of 1 gram of water from 0°C to 100°C?

To do this last calculation, we need to know what's called the specific heat of water — which is defined as the amount of heat it takes to raise the temperature of 1 *kilogram* of water by one degree Celsius. Its value is

Specific heat of water $= 4.190 \times 10^{3}$ J/kg degree.

(Don't get the idea that scientists store all these basic numbers in their heads. They don't. They carry around a little booklet or a smartphone that's got everything listed there. No need to remember all that stuff.)

So finally, the amount of heat needed to raise the temperature of 1 gram of water from 0°C to 100°C is given by

(mass of water)(sp ht of water)(change in temp)

$$= (10^{-3}\,\text{kg})(4.190 \times 10^3\,\text{J/kg degree})(100\,\text{degrees})$$

$$= 419\,\text{J}.$$

Admittedly, the 102 J of heat released per hour by 1 gram of radium-226 is less than the 419 J of heat needed to raise the temperature of 1 gram of water from freezing to boiling. But it's in the right ballpark. Doesn't detract from Curie and Laborde's main point — radioactive decay generates heat. A lot of it.

Need to allow for that in any discussion of the age of the Earth. Exit Kelvin.

Bibliography

There are many books, articles, and websites that tell the story of radioactivity. Those that I used were the following:

Gorst, M. 2001. *Measuring Eternity: The Search for the Beginning of Time.* New York: Random House Inc.

Burchfield, J.D. 1990. *Lord Kelvin and the Age of the Earth.* Chicago: University of Chicago Press.

Jackson, P.W. 2006. *The Chronologers' Quest: The Search for the Age of the Earth.* Cambridge: Cambridge University Press.

Books that contain more technical detail are the following:

Brush, S.G. 1996. *A History of Modern Planetary Physics, Vol 2: Transmuted Past.* Cambridge: Cambridge University Press.

Stacey, F.D. 1969. *Physics of the Earth.* New York: John Wiley and Sons, Inc.

Dalrymple, G.B. 1991. *The Age of the Earth.* Redwood City: Stanford University Press.

Stephen Brush has also responded to some religious issues in

Brush, S.G. 1982. Finding the Age of the Earth by Physics or by Faith. *Journal of Geological Education* **30**, 34–58.

The research of Clair Patterson and colleagues is described in

Patterson C., Tilton G., and Inghram M. 1955. Age of the Earth. *Science* **121**, 69–75.

Chapter 6

Concluding Remarks

So, which is it? Six thousand years? Or four and a half billion years?

As we have seen, the 6000 years is determined primarily by taking numbers from the Bible. The King James Version. Based on two-thousand-year-old texts. Following the ages of the patriarchs and the events described therein. Not questioning the numbers themselves. (Well, hardly ever.) But that took us only so far. Then what? Follow Ussher. Use "*all* History, as well Sacred, as Prophane". And "methodically digest" it. Brought us to Nebuchadnezzar. Died 3442 years after Creation in the biblical chronology; 562 BC according to the archaeologists. From which Ussher concluded that the "Let there be light" had happened in 4004 BC. Spoken by a Creator. The gospel truth to some. (Though not part of the Gospel itself.)

How does it compare with what we see in the world around us ... using — as some might say — our God-given intelligence?

James Hutton couldn't buy the biblical creation story — at least not in its entirety. The *rocks* to him were God's book of Nature, written by "God's own finger". Told a different story. Couldn't all have happened in one week. Or caused by one deluge that had lasted only forty days and forty nights. Instead, he saw repeated cycles of erosion, sedimentation, compactification, and uplift. All of which demanded eons of time. Driven by energy from within. That, he said, was the message of the rocks.

Substantiated eventually by methods based on radioactivity. That vertical greywacke at Siccar Point? 430 million years old. Formed during what geologists nowadays call the Silurian Period (though the Celtic tribe the Silures were not around at the time). And the horizontal red sandstone? Only 380 million years old. In the Devonian Period. Younger, as Hutton said.

And what about all that upheaval at the surface of the Earth? Nowadays it's understood in terms of tectonic plates. Crashing into one another. Admittedly at a rate of only a centimeter or so per year. Not exactly break-neck speed. But sufficient to cause land to go up (think Himalayas) or go down (think Mariana Trench). Energy supplied by the movement of the hot molten innards of the Earth.

Charles Darwin liked the idea of an ancient Earth. It provided plenty of time for his evolution-by-natural-selection to work its magic. At a snail's pace. Survival of the fittest. Soon to be developed further. For, as Darwin was busy writing his *Origin of Species*, an unknown Augustinian friar by name of Gregor Mendel in far-off Brno (in what is now the Czech Republic) was busy discovering the *quantitative* laws of inheritance. By studying peas. Couldn't get more basic than that. Eventually blossomed in the 20th century into genetics. And DNA. Which provides today's understanding of the origin of species. In detail.

Lord Kelvin disagreed with an ancient Earth. Could prove it, he said. By injecting physics. The conservation of energy accompanied by the irreversible dissipation of heat. Decided that the Earth had taken only about a hundred million years to cool down to its present state. He had calculated it. Incontrovertible. Till it was proved wrong. Destroyed by convection. And radioactivity.

Which brought us the confirmation of the atom and the discovery of the nucleus. In various flavors. With specific decay chains. Analyze many different rocks. From both above and below. Using many different methods. Results all consistent. Gives a specific number for the age of the solar system. Which includes the Earth. About four and a half billion years.

So, take your pick. Which is it? The contrast couldn't be greater. Made even more so if you look at the specifics.

The biblical creation story

In Genesis 1, underneath the 4004 BC date printed at the top of the page, Creation took place in only six days.

Day 1. The Heavens and the Earth created. The Earth formless, void, covered with water, pitch-dark. Then primordial light, bringing day and night.

Day 2. The waters separate — above the sky and below the sky.

Day 3. The waters under the sky separate further, resulting in dry land and seas. Followed by vegetation covering the dry land.

Day 4. The Sun, Moon, and stars are created.

Day 5. Creatures populate the sea, and birds the air.

Day 6. Wild animals, cattle, and creeping creatures appear. Finally, mankind — male and female.

An amazing story. The cosmos all created by one designer, Yahweh. Followed by the story of the relationship between this Creator and his human creations.

The scientific creation story

A very different story. Mapped out by modern scientists. From observing, measuring, deducing, testing, predicting, and confirming.

Stage 1. Spacetime. Wrapped in a point. Huge energy burst. Maybe a quantum fluctuation. The beginning of the universe. 13.8 billion years ago. Misnamed "Big Bang" — no one there to hear it. Extreme heat. Temperature begins to plummet.

Stage 2. Continued rapid expansion. Exponential. But very brief. Called "inflation". The fundamental forces of Nature emerge. In quantum form.

Stage 3. Rate of expansion slows. The building blocks of matter begin to appear. Such as electrons. And quarks. Which soon coalesce into protons and neutrons. Immersed in a cloud of photons. Temperature continues to drop. Nuclei appear. It's now about 3 minutes after the Big Bang. Later, at about 380,000 years, it's cool enough to allow atoms to form, the photons in the expanding space–time no longer having enough energy to break them apart, are liberated.

Stage 4. Myriads of galaxies begin to form. With countless numbers of stars. Which themselves evolve. Gravity and nuclear fusion at work. Generating the elements. Sometimes exploding as spectacular supernovae.

Stage 5. A major change begins. The rate of space–time expansion starts to increase. Accelerates. And at about the same time, a major event — the appearance of our very own Milky Way Galaxy. Another dubious name — "Galaxy" itself means "Milky". A tautology. Began to shine about five or so billion years ago. And near its edge, our Sun. With its companion planets. Including ours.

Stage 6. Planet Earth. Our eventual home. Hot, but cooling down. Developing a solid core. Surrounded by molten rock. At the surface, a mosaic of plates. Supporting continents, and oceans. Pushing and shoving. Causing earthquakes and volcanoes. Eventually life appears. Microscopic at first. Evolving into all the life forms we know today. Including *Homo Sapiens*. About two million years ago. Our ancestors.

That's the science creation story. Omitting the mathematics. A great achievement. An apple in the eye of most beholders. Figured out from countless observations and measurements. Based on the laws of physics — God-given, some would say. Can even make predictions. Such as the Cosmic Microwave Background — those photons that escaped entrapment after they were no longer able to break up atoms. Finally, detected in 1954. Prediction. Confirmation. The essence of science. Can't do that in non-science. And not yet a finished product. Still got a ways to go. Like finding out what most of the universe is made of! Dark matter and dark energy. Exciting times ahead! What will be the final design?

All of which doesn't of course answer the Big Question for scientists, and everyone else — was there a Guiding Hand? That's a tougher question. Personal. Calls for an odyssey of a different kind. For each one of us.

Index